Cloning

For and Against

Edited by

M.L. Rantala and Arthur J. Milgram, Ph.D.

OPEN COURT
Chicago and La Salle, Illinois

To order books from Open Court, call toll-free 1-800-815-2280

Cover photograph of two cloned sheep, Megan and Morag:
Spooner/The Gamma Liaison Network. Used by permission.

Volume 3 in the series 'For and Against'

First printing 1999

Printed and bound in the United States of America.

Library of Congress Cataloging-in-Publication Data

Cloning : for and against / edited by M.L. Rantala and Arthur J. Milgram
p. cm.
Includes bibliographical references and index.
ISBN 0-8126-9374-4 (hbk. : alk. paper)
ISBN 0-8126-9375-2 (pbk. : alk. paper)
1. Human cloning--moral and ethical aspects--Miscellanea. 2. Human cloning--Philosophy--Miscellanea. I. Rantala, M. L. II. Milgram, Arthur J., 1964-
QH422.2 .C564 1999
176--dc21

98-50289
CIP

To Gregor Mendel (1822–1884),
and the priceless complexity
and beauty of all life.

Contents

Permissions Acknowledgments

The editors gratefully acknowledge permission from the following copyright holders to reprint many of the readings in this book:

BioScience and Beth Baker for "To Clone or Not to Clone" (© American Institute of Biological Sciences 1997); *British Medical Journal* for "The Promise of Cloning for Human Medicine" (March 1997); *Canada and the World* for "Ethical Debate" (January 1994); Catholic News Service, for "Criticism of Cloning Mounts," which appeared in the *National Catholic Reporter* (23 January 1998); *Christian Century* for "To Clone or Not to Clone?" © 1997 Christian Century Foundation. Reprinted by permission from the March 1997 issue of the Christian Century; *Christianity Today* and John Kilner for "Stop Cloning Around" (28 April 1997); *Chronicle of Higher Education* and Lori B. Andrews for "Human Cloning: Assessing the Ethical and Legal Quandaries," and *Chronicle of Higher Education* for "Senate Rejects Bill to Ban Human Cloning" and "Debate Over Cloning Touches One of Society's Most Sensitive Nerves" (all three February 1998); *Commonweal* for "The Mystery Remains," "Cloning Isn't Sexy" (both 28 March 1998); *Discover* magazine for "Fear and Longing" (1998), "Splitting Heirs" (January 1994), "A Sheep in Sheep's Clothing" (January 1998), "A Clone of One's Own" (1998); *E: The Environmental Magazine* for "Me and My Shadow" (July–August 1997), P.O. Box 2047, Marion, OH 43306. Telephone: 815-734-1242. Subscriptions are $20 per year; *Fortune* magazine, for "The Real Biotech Revolution" (*Fortune,* 31 March 1997); *The Futurist* for "Will Cloning End Human Evolution?"; *Hastings Center Report* for "Cloning: The Work Not Done," "Ban Cloning? Why NBAC Is Wrong" (September–October 1997); *Humanist* for "Second Thoughts about Cloning Humans" (May–June 1997); *Maclean's* for "Send in the Clones" (18 March 1996), "The Prospect of Evil," "Body Doubles" (both 10 March 1997); *National Review,* 215 Lexington Avenue, New York, NY 10016 for "All the Same" (30 June 1997), "Seeds of Trouble" (9 February 1998), "Cloning Cloning Cloning" (24 March 1998); *New Leader* for "Between Issues" (24 February 1997); *New Republic* for "Bad Seed" (9 February 1998); *New Statesman* for "Human Genetics: The Dinner-Party Guide" (© New Statesman 1998); *Politics and the Life Sciences* for "Creating a Clone in Ninety Days: In Search of a Cloning Policy" (September 1997); Rowman and Littlefield for an excerpt from Gregory E. Pence, *Who's Afraid of Human Cloning?* (Lanham, Md: Rowman and Littlefield, 1998), pp. 119–122; Stephen Garrard Post and America Press, Inc., 106 West 56th Street, New York, NY 10019, for "The Judeo-Christian Case against Human Cloning" *(America,* June 1997); *Reason* for "The Twin Paradox" (May 1997), "Fatalist Attraction" (July 1997), "Clone Wars" (1998); *Science* for "After Dolly: A Pharming Frenzy" (30 January 1998), "Biomedical Groups Derail Fast-Track Anti-Cloning Bill" (20 February 1998); *Science News* for "Gene Cloned for Stretchiest Spider Silk" (21 February 1998); Time Life Syndication for the following articles from *Time:* "A Special Report on Human Cloning" (10 March 1997), "Will We Follow the Sheep?" (10 March 1997), "Can Souls Be Xeroxed?" (10 March 1997), "Of Headless Mice and Men" (19 January 1998), "The Case for Cloning" (9 February 1998); *U.S. News and World Report* for "The World after Cloning" and "Human Cloning? Don't Just Say No" (© U.S. News & World Report, 1997).

Preface

The debate over cloning, especially possible future cloning of human beings, is one of the most passionate and divisive controversies of our time. This book contains a fair, balanced selection of short writings taking differing views on various aspects of the cloning issue.

We hope that this selection will be used in classes to stimulate debate, and will also appeal to general readers interested in this highly controversial area. Both the titles of the readings and most of the sub-headings were not part of the original articles or excerpts, but were added by the editors. Our Introduction explains some of the scientific background to the cloning controversy, as a guide to the reader.

Several people generously aided us in the formation of this collection. Sharin Bowers, Lourie Wilson Reichenberg, and Barbara Garmon provided detailed and immensely efficient help in securing reprint permissions.

Dr. Tatyana Grushko, Dr. Margaret Morgan, Joy Rose, and Michael Marsh assisted in various important ways.

Many thanks to Kerri Mommer, Jeanne Kerl, and Jennifer Asmuth at Open Court for their encouragement and patience. David Ramsay Steele, editorial director at Open Court, was an invaluable help at all stages of the work and was never without the right answer to even the trickiest questions.

As always, we are grateful for the support of our parents and indebted to Ilma, Ruma, and Harmaa for the joy they bring us each and every day.

M.L. Rantala,
Chicago, Illinois

Arthur J. Milgram
Philadelphia, Pennsylvania

Editors' Introduction

The Scientific Background

Scottish scientist Ian Wilmut inaugurated a new chapter in science when he introduced the world to Dolly the lamb. To look at her, Dolly seemed the quintessential sheep: woolly, imperturbable, and utterly indistinguishable from thousands upon thousands of others in her species.

But hers was a difference of momumental proportions. She was a clone of a six-year-old Finn Dorset ewe. Dolly was the first mammal to be cloned from adult body cells.[1] Judging from the public reaction Dolly has aroused, she is almost the only one who hasn't found the distinction fascinating, frightening, exciting, or execrable.

Most people wouldn't notice the difference if the world's sheep farms were populated with clones. But what about human cloning? Wouldn't you notice if your boss cloned herself for every one of the firm's managerial positions?

It is the possibility of cloning humans which has captured attention all over the world. Will we see a world where Frankenstein monsters turn on their creators? Are those opposed to cloning failing to consider its beneficial effects?

Rapid and widely disseminated results in science have increased public awareness of scientific research. Scientists were once perceived as an elite group of eggheads, but now they are seen as human just like the rest of us. Science has become more important to society, and people are trying harder to understand it, and to use it wisely and rationally.

An Earlier Biological Controversy

The current controversy surrounding animal cloning and the possibility of human cloning does not mark the first time that biological innovation has generated such widespread concern. With the advent of recombinant DNA in the late 1960s and early 1970s came fears about genetically altering bacteria. "Recombinant DNA" refers to cloning genes and the ability to transfer those genes between species and individuals. The thought of inserting human genes, such as the gene for insulin, into a bacterium resulted in a public outcry, calls for outright bans on such techniques, and city, state and federal attempts at legislating the activities involved in recombinant DNA research.

In the early 1970s a moratorium on recombinant DNA techniques was voluntarily imposed by scientists in the U.S. so that such issues could be publicly discussed before proceeding with the full scale application of this new scientific technique.[2]

Tool of the Elite or Misplaced Fears?

Soon molecular biology was accused of being a tool of the rich industrial elite to further their own commercial interests.[3] Genetic screening of fetuses *in utero* was claimed to be a form of human genetic engineering much like eugenics,[4] there was even fear that *in vitro* fertilization—growth of embryos in test tubes—opened the door to human cloning. Some thought that scientists wanted to create human clones to do their laboratory work—ignoring the fact that a ready supply of such creatures already exists; they are called graduate students.

None of the fears expressed about recombinant DNA has materialized. Guidelines and regulations for conducting recombinant DNA research have been developed.[5] More recently, much of the authority and responsibility for review of human gene transfer research has been transferred from the NIH's Recombinant Advisory Committee to the Food and Drug Administration.[6]

Falsely Claiming a Clone: The Rorvik Hoax

In His Image: The Cloning of a Man by David Rorvik (1978) claimed to be a true account of the actual creation of a human clone. There ensued a hearing before the congressional subcommittee on Health and the Environment in May of 1978. There was considerable discussion of the view that scientists—biologists in particular—were engaging in a conspiracy to develop the technology necessary and actually to clone human beings.

A representative from California, Henry Waxman, asserted that cloning as he understood it could someday be used to create a "super race," something which Alvin Toffler in *Future Shock* referred to as a biological Hiroshima: "The question is not, can new life survive outside the laboratory, but can our traditional values survive within them?"[7] The ethical, legal, and social implications of work in biology at that time bear a strange similarity to the current discussion concerning what is now the reality of mammalian cloning.

Science Fiction versus Science Fact

The scientists who were called to testify regretted the fact that this whole discussion of cloning was provoked by science fiction rather than real science, and emphasized that the work they were doing did not relate to reproductive biology. They feared that misunderstanding of the situation could lead to government restrictions and efforts to control problems which didn't exist.

As with many scientific discoveries of the twentieth century, the advances in molecular biology achieved over the last thirty years have, and will continue to have, significant social impact. The rapid pace of scientific progress in general has dramatically altered the lives of most of us—often in ways we do not even recognize or have come to accept with almost equal rapidity. After all, who has not used an automated teller machine, or how many of our young tackle their homework with the aid of an electronic calculator, or even a computer?

The Human Genome Project

The Human Genome Project is a multinational scientific collaboration with the purpose of determining the DNA sequence of the entire human genome. (A genome is the sum total of all the DNA of a living being.) The project holds great potential, but the knowledge it brings may entail significant risks. Among the many promises held out by the Human Genome Project is the ability to identify traits which may be linked to heritable disorders. This, it is hoped, will allow for targeted research into the biochemical causes of heritable and related diseases and provide treatments and therapies more rapidly than conventional research approaches have been able to provide up until now. Cloning will be an important element in many of the new approaches being developed.

However, social implications caused by the knowledge the genome brings can be anticipated before completion of the fundamental research:

> [T]he ability to identify people who have or are carrying genetic diseases inevitably comes before treatments are discovered, as a result, when all you can do is diagnose and you can't treat, each success in the gene-mapping project will create new ethical, legal and social implications—or ELSI in government jargon.[8]

This is only one of the many social implications of the research being done.

The Human Genome Project was initiated in the 1980s by the Department of Energy (DOE) and is currently a joint project between the DOE and the National Institutes of Health (NIH), but it is also international in scope. When the Human Genome Project was started, the technology required to accurately and completely sequence the human genome did not exist. Accordingly, technology development to more rapidly and cost effectively accomplish the Human Genome Project's goal of sequencing all 24 human chromosomes was a significant component of the Project.

The advances in technology have been sufficient to allow the establishment, around the world, of genome projects for several agricultural crops, and the DOE began the Microbial Genome Initiative to characterize the genomes of microbes which are "... of interest from an energy, environmental, or industrial perspective."[9] Among the applications which will be affected by these scientific efforts are the production of energy, monitoring and restoration of environmental quality, agricultural and materials production, and industrial processes such as chemical conversions and fiber production. One aspect of expected scientific advances that can be attributed to the Genome Project is that the resources developed and new approaches and technologies that issue from it will enable new fundamental research and applications. In other words, the benefits incurred from this research are not necessarily the data itself, or the direct outcomes of the work.

The stakes involved in this scientific endeavor are tremendous. Commercial interests in genomic research has come from the agricultural chemical and pharmaceutical industires. To date several agricultural crop genomes are nearer completion than the human genome project. A multitude of bacterial genomes have already been sequenced, and many more will eventually be finished. Industries are spending tremendous amounts of money simply to patent (and thus gain commercial control over) individual genes and whole genomes.[10]

One might suspect that the human genome project will facilitate the creation of genetically engineered cloned people. After all, if we were to clone a person, and we had the complete sequence of that person's DNA, we might also consider making purposeful changes to that DNA before proceding with the cloning. The creation of genetically engineered livestock was a motivation of the original Roslin, Scotland experiment which created Dolly.

Just What *Is* a clone?

With much research being done to map out our genes and to advance cloning techniques, it is natural to ask, just what is a clone, anyway?

A clone can mean many different things in biological research depending on the context in which it is used. At the molecular and cellular level, cloning has been conducted by biologists for several decades. Individual genes are commonly cloned, as are single human, plant, and animal cells. Under limited circumstances entire organisms can be cloned. When an entire organism is cloned, the result is a genetically identical copy of the original organism (the precise meaning of genetically identical is explained below).

Clones or genetically identical copies of whole organisms are commonplace in the natural world, and cloning has been in use in the plant breeding community for centuries. Genetically identical cloned plants are commonly referred to as "varieties" rather than "clones."[11]

Each of Us Is Unique

Identical twins are more closely identical to each other than even Dolly is to her genetic parent. This is because in addition to our nuclear DNA, there are small amounts of DNAs contained outside of the nucleus of each cell. Since Dolly was formed by a nuclear transplantation of a parental nucleus into a donor egg, she inherited the DNAs outside of her nucleus from the donor egg. These DNAs differ from her genetic parent. Thus, identical twins are more truly clones than even Dolly[12] for their DNA is identical both inside and outside the nucleus. Yet who can fail to notice the extensive differences in personality between identical twins? Even identical twins raised together in the same home are without question very distinct individuals. If one of an identical twin embryo were to be frozen, and then implanted in a surrogate mother's womb at a later point in time, would it be reasonable to expect this twin to be any more similar to its identical sibling born earlier than a pair of such twins born to the same parents and raised together?[13]

So each of us is and must be unique. This is true no matter how similar our genes may be to any sibling or clone. The debates may rage about the proportion of environment and the proportion of genetics required to make us who we are. But even when the debates are over forever, we will never be able to duplicate the precise conditions, both genetic and environmental to create another you.

What Came Before Identical Sheep?

Cloning of individual cells is a method used to grow cells in petri dishes. Vegetative or somatic cells—cells which are not reproductive cells such as sperm and eggs—can be isolated from an organism and maintained in a liquid media culture. Cells derived from most plants can be maintained indefinitely in such cultures, and new plants can be regenerated from each single cell any time one wishes. However, cells from most animals will only divide a limited number of times in culture, and cannot be maintained ordinarily for long periods of time.[14] Cells isolated and grown in culture yield cells which are genetically identical clones of each other. Cell cloning and tissue culture allows for the production of large amounts of genetically identical cells, which can be used for genetic and molecular experiments that cannot be conducted on only single cells. Such techniques are instrumental in the testing and development of new medicines.

Genetic identity is a reference to the DNA contained within each cell. DNA is a linear molecule which in many ways is similar to a string. A single strand of DNA consists of chemicals called nucleotides (commonly referred to as bases in this context) strung together one after the other. However, DNA is found in the form of a double stranded helix, in which two single strands of DNA wrap around and adhere to each other. The two single strands of DNA are joined to each other through what is called basepairing.

Basepairing refers to the capability of the different nucleotides of DNA to bind to their complementary nucleotides on the other strand of DNA. Nucleotides come in four flavors for DNA. The precise names are only important to biochemists, and the four bases are commonly abbreviated A, T, G, and C. The order of these nucleotides along the strand of DNA is the so-called DNA sequence. In the double-stranded helix in which two strands of DNA are held together within cells, each nucleotide is capable of forming bonds with a complementary nucleotide on the opposite strand. For example, A binds specifically with a T on the opposite strand of DNA and G binds with its complementary base: C. Thus, the two stands of DNA in the double helix hold on to each other kind of like a zipper when it is closed, where the individual bases would correspond to the teeth of the zipper.

Chromosomes: The Package for DNA

In humans (and all other animals and plants) our DNA is packaged in the form of chromosomes. Chromosomes are long strands of a single molecule of double-stranded DNA. Human beings have a total of 46 such chromosomes in each nucleus of every cell in the body. The 46 chromosomes together contain some 3 billion basepairs, which means that the human DNA sequence is 3 billion nucleotides long. This 3 billion basepair sequence is equal to one genome. If all the chromosomes were removed from the nucleus and the DNA was stretched out straight, and placed end to end, it would measure about one meter in length. We have within our bodies enough cells that if all the DNA in our bodies were stretched end to end, it would reach to the sun and back.

For two organisms to be genetically identical means that the the order of the nucleotides strung along each chromosome contained in each cell from both organisms is exactly the same. In the case of humans, such as identical twins, that would be two genomes of 3 billion bases each having the same sequence of nucleotides from start to finish.

Genes are the Information of Life

Genes are regions of DNA on a chromosome which contain specific kinds of informatin within the DNA sequence. Genes direct the synthesis of proteins. Proteins are basically the stuff that living things are made out of. Genes of particular interest have been routinely cloned for almost 30 years, using recombinant DNA techniques. Genes can be isolated and cloned from almost any organism's DNA.

In the science of molecular biology an animal clone refers to genetically identical animals derived by means of non-sexual reproduction of a common ancestor. This is to say that individuals (be they single cells or whole organisms) which are brought to the light of Earth by a single parent without the means of any genetic rearrangement are genetically identical to this parent, they are therefore clones. To the lay person, this is not really an answer to the question: "what is a clone?"

Most animal and plant cells contain at least two duplicate sets of chromosomes. One set is of maternal origin, the other is paternal. That is, there are two copies of each chromosome. Thus, as mentioned above, humans have 46 chromosomes, 23 are from the mother, and the other 23 are from the father. Each human cell has two sets of 23 chromosomes (for example we all have two copies of the single human chromosome 22).[15] One copy of each chromosome pair is derived from one or other parent.

In ordinary cell division the replicated chromosomes are simply divided evenly during cell division to produce two new cells. But in recombination the maternal and paternal[16] chromosomes within the cell actually exchange DNA with each other. Essentially by rearranging the DNA between each parental chromosome new combinations of genes not found in either parent are created.

Plants Are Not Like Birds and Bees

Sexual reproduction as it occurs in humans is almost an unusual case among the many different options living things use to reproduce themselves. The plant world, unlike the birds or the bees, provides many comprehensible examples of sexual reproduction as well as natural instances of cloning.

Plants as we all understand them flower, and after pollination go to seed. The next generation of plant then grows from the seed. This is the general paradigm used by most plants that most people are familiar with (the case of mosses or ferns is slightly more complex). Within the plant flower, gametes are formed, and mating takes place.

Some plant species are capable of fertilizing themselves. This would be akin to single humans being capable of producing both sperm and eggs. How much simpler relations between the sexes might be, if we could each produce our own offspring!

There are also species of fungi (many placed in the Fungi Imperfecti) in which sexual reproduction is not known, and only clonal reproduction occurs.[17] In worms, such as the flatworm *Planaria*, any part of the worm can regenerate an entire worm. Thus, the worm can be cloned from any part of itself. In higher animals such as insects missing parts can at least be regenerated even though the organism itself cannot be cloned from a mere part.

Many plants are capable of cloning themselves. Clonal reproduction is a widespread characteristic among the hepatophyta and bryophyta (liverworts and mosses). The liverwort *Marchantia*, for example, produces specialized structures called gemmae. These are small lens-shaped bodies produced on the upper surface of the plants which consist of vegetative tissue. Gemmae are capable of regenerating new plants when removed from the parent and placed in conditions favorable to growth. This principle is similar to the fragmentation of certain species of algae or to vegetative propagation of vascular plants, for example when we root rooting a cutting taken from a favorite houseplant, and to the regeneration of worms from fragments of themselves. The regenerated plants are genetically identical to their single parent. This is because vegetative, or in general non-reproductive, tissue was caused to regenerate an entire plant without the intervention of sex.

Sex Is Recombination–Cloning Is Simple Replication

As we have seen, a fundamental outcome of sexual reproduction is the recombination of genetic material. In cloning, by contrast, there is no recombination, only replication. The DNA is simply copied without changes. Cloning is a fundamental method of reproduction in large parts of the natural world. The ability of an organism to be cloned or to regenerate missing parts is of fundamental importance to biological theory.

But cloning mammals? Cloning transgenic creatures designed to 'grow' material for human transplantation? Cloning a human being? Does cloning mean that men and sex are obsolete? Is cloning a way to save endangered species like the hairy-nosed wombat or will it destroy genetic diversity? Must we ban cloning? Can we ban cloning? These are just a few of the questions which follow the introduction of the woolly, imperturbable lamb, Dolly. These are some of the questions that this book will enable the reader to begin to explore.

Notes to the Editors' Introduction

1. Dolly was named after Dolly Parton. The adult cell from which she was cloned came from mammary tissue.

A technical account of how Dolly was cloned can be found in I. Wilmut, A.E. Schnieke, J. McWhir, A.J. Kind, and K.H.S. Campbell, "Viable Offspring Derived from Fetal and Adult Mammalian Cells," *Nature*, v385, February 27, 1997, pp.810–813.

2. See *Research with Recombinant DNA* (National Academy of Sciences, Washington D.C. 1977) for a discussion of the

concerns at that time, and an overview of events surounding the voluntary moratorium. See also some of the readings in this volume.

3. Susan Dawkins. *Gene Wars: The Politics of Biotechnology*, (New York: Seven Stories Press, 1997).

4. See Jonathan Beckwith's essay, reprinted in this volume (Reading 47) as, "Cloning Serves the Interests of Those in Power".

5. See for example "Guidelines for Research Involving Recombinant DNA Molecules." (Department of Health and Human Services, National Institutes of Health, June 1994.)

6. Some discussion of motivations and public concerns regarding changing the role and authority of the Recombinant Advisory Committee are found in the *Federal Register* (vol. 62 no, 31. p. 7108, 1997).

7. Subcommittee on Health and the Environment of the Committee on Interstate and Foreign Commerce, House of Representatives, Washington D.C. 1978.

8. Opening statement of Harris W. Falwell. Hearing before the Subcommittee on Energy of the Committee on Science, Space, and Technology, House of Representatives, October 4, 1994.

9. Statement of Dr. Ari Patrinos. Hearing before the Subcommittee on Energy of the Committee on Science, Space, and Technology House of Representatives, October 4, 1994.

10. A very interesting essay on the negative social effects of this "gene race" can be found in *Gene Wars,* reference above.

11. "The Science and Application of Cloning" in *Cloning Human Beings Report and Recommendations of the National Bioethics Advisory Commision*, 1997.

12. Testimony of Dr. Harold Varmus, Nobel laureate and director of the National Institutes of Health, before the Subcommittee on Public Health and Safety of the Committee on Labor and Human Resources, United States Senate, 1997. *Examining Scientific Disoveries in Cloning, Focusing on Challenges for Public Policy.* U.S. Government Printing Office, Washington D.C.

13. This example is based on a discussion by Stephen Jay Gould in "Dolly's Fashion and Louis's Passion," in *Natural*

History, June 1997, reprinted in *Clones and Clones*, edited by Martha C. Nussbaum and Cass R. Sunstein (New York, Norton, 1998).

14. Such cell lines are said to be mortal. Individual cells can be "immortalized" by treatment with chemicals called teratogens (a subset of chemicals known as mutagens). Immortalized cells have been "transformed" by the teratogens, and can be maintained indefinitely in culture. By studying the genetic changes induced in such transformed cells, the discovery of genes which are involved in the regulation of cell growth. These genes are called oncogenes, and mutations in them leads to unrestricted cell growth. Mutations in oncogenes of somatal tissue of adult animals is believed to be one cause of cancer.

15. The human chromosomes are visual during cell division, and each chromosome has a distinctive appearance allowing the specific chromosomes to be distinguished from each other. This has allowed names to be given to each chromosome. Thus, we have two pairs of each of 23 different chromosomes. They are named chromosomes 1, 2, 3, ... , 22, and the 23rd chromosome is either X or Y.

16. In this context the mternal and paternal chromosomes are often refered to as homologous chromosomes. Homologous chromosomes are essentially just chromosomes with the same names. The chromosome 22 of maternal origin, and the chromosome 22 of paternal origin are for example homologous.

17. Robert F. Scagel *et al. Plant Diversity: An Evolutionary Approach,* (California: Wadsworth Publishing Co. 1969).

I

What Is Cloning?

Introduction to Part I

Although there had been several previous waves of interest in cloning, reports in 1997 that Dr. Ian Wilmut had cloned an adult sheep from one of its adult cells unleashed a storm of new controversy. Often, discussions and fictional portrayals in the popular media showed that the notion of a clone was badly misunderstood. The most serious misunderstanding was the belief that a human clone would not be a separate, independent, individual human being. A clone is not an extension of its 'parent'; the clone does not share the parent's thoughts or memories. A clone and its parent would be genetically identical, as identical twins are, but no more than that.

Reading 1, by Herbert, Sheler, and Watson, gives straightforward answers to some basic questions about cloning and its reception by religious leaders. Robert Wright (Reading 2) raises questions about the extent to which who we are is influenced by our genes, and probes the extent to which we might enjoy an "uncanny empathy" with someone who has all our genes. In Reading 3, Marc Zabludoff explains why we can never produce an exact duplicate of ourselves.

Josie Glausiusz (Reading 4) helps us understand the distinctive nature of cloning by explaining how it differs from the embryo splitting accomplished by Hall and Stillman. In Reading 5, Vincent Kiernan introduces us to one of the ethical issues linking cloning with abortion and other controversial topics: is a cloned cell or group of cells an embryo? Finally, Caroline Daniel presents some basic facts of current work on human genetics.

1

A Clone Would Have a Soul

Wray Herbert, Jeffery L. Sheler, and Traci Watson

At first it was just plain startling. Word from Scotland last week that a scientist named Ian Wilmut had succeeded in cloning an adult mammal—a feat long thought impossible—caught the imagination of even the most jaded technophobe. The laboratory process that produced Dolly, an unremarkable-looking sheep, theoretically would work for humans as well. A world of clones and drones, of *The Boys from Brazil* and *Multiplicity*, was suddenly within reach. It was science fiction come to life. And scary science fiction at that.

In the wake of Wilmut's shocker, governments scurried to formulate guidelines for the unknown, a future filled with mind-boggling possibilities. The Vatican called for a worldwide ban on human cloning. President Clinton ordered a national commission to study the legal and ethical implications. Leaders in Europe, where most nations already prohibit human cloning, began examining the moral ramifications of cloning other species.

Like the splitting of the atom, the first space flight, and the discovery of "life" on Mars, Dolly's debut has generated a long list of difficult puzzles for scientists and politicians, philosophers and theologians. And at dinner tables and office coolers, in bars and on street corners, the development of wild scenarios spun from the birth of a simple sheep has only just begun. *U.S. News* sought answers from experts to the most intriguing and frequently asked questions.

Why would anyone want to clone a human being in the first place?

The human cloning scenarios that ethicists ponder most frequently fall into two broad categories: 1) parents who want to clone a child, either to provide transplants for a dying child or to replace that child, and 2) adults who for a variety of reasons might

---

This appeared as "The World After Cloning" in *U.S. News and World Report*. v122 (Mar. 10, 1997). pp. 59–63.

want to clone themselves.

Many ethicists, however, believe that after the initial period of uproar, there won't be much interest in cloning humans. Making copies, they say, pales next to the wonder of creating a unique human being the old-fashioned way.

Could a human being be cloned today? What about other animals?

It would take years of trial and error before cloning could be applied successfully to other mammals. For example, scientists will need to find out if the donor egg is best used when it is resting quietly or when it is growing.

Will it be possible to clone the dead?

Perhaps, if the body is fresh, says Randall Prather, a cloning expert at the University of Missouri—Columbia. The cloning method used by Wilmut's lab requires fusing an egg cell with the cell containing the donor's DNA. And that means the donor cell must have an intact membrane around its DNA. The membrane starts to fall apart after death, as does DNA. But, yes, in theory at least it might be possible.

Can I set up my own cloning lab?

Yes, but maybe you'd better think twice. All the necessary chemicals and equipment are easily available and relatively low-tech. But out-of-pocket costs would run $100,000 or more, and that doesn't cover the pay for a skilled developmental biologist. The lowest-priced of these scientists, straight out of graduate school, makes about $40,000 a year. If you tried to grow the cloned embryos to maturity, you'd encounter other difficulties. The Scottish team implanted 29 very young clones in 13 ewes, but only one grew into a live lamb. So if you plan to clone Fluffy, buy enough cat food for a host of surrogate mothers.

Would a cloned human be identical to the original?

Identical genes don't produce identical people, as anyone acquainted with identical twins can tell you. In fact, twins are more alike than clones would be, since they have at least shared the uterine environment, are usually raised in the same family, and so forth. Parents could clone a second child who eerily resembled their first in appearance, but all the evidence suggests the two would have very different personalities. Twins separated at birth do sometimes share quirks of personality, but such quirks in a cloned son or daughter would be haunting reminders of the child who was lost—and the failure to re-create that child.

Even biologically, a clone would not be identical to the

"master copy." The clone's cells, for example, would have energy-processing machinery (mitochondria) that came from the egg donor, not from the nucleus donor. But most of the physical differences between originals and copies wouldn't be detectable without a molecular-biology lab. The one possible exception is fertility. Wilmut and his coworkers are not sure that Dolly will be able to have lambs. They will try to find out once she's old enough to breed.

Will a cloned animal die sooner or have other problems because its DNA is older?

Scientists don't know. For complex biological reasons, creating a clone from an older animal differs from breeding an older animal in the usual way. So clones of adults probably wouldn't risk the same birth defects as the offspring of older women, for example. But the age of the DNA used for the clone still might affect life span. The Scottish scientists will monitor how gracefully Dolly ages.

What if parents decided to clone a child in order to harvest organs?

Most experts agree that it would be psychologically harmful if a child sensed he had been brought into the world simply as a commodity. But some parents already conceive second children with nonfatal bone marrow transplants in mind, and many ethicists do not oppose this. Cloning would increase the chances for a biological match from 25 percent to nearly 100 percent.

If cloned animals could be used as organ donors, we wouldn't have to worry about cloning twins for transplants. Pigs, for example, have organs similar in size to humans'. But the human immune system attacks and destroys tissue from other species. To get around that, the Connecticut biotech company Alexion Pharmaceuticals Inc. is trying to alter the pig's genetic codes to prevent rejection. If Alexion succeeds, it may be more efficient to mass-produce porcine organ donors by cloning than by current methods, in which researchers inject pig embryos with human genes and hope the genes get incorporated into the embryo's DNA.

Wouldn't it be strange for a cloned twin to be several years younger than his or her sibling?

When the National Advisory Board on Ethics in Reproduction studied a different kind of cloning a few years ago, its members split on the issue of cloned twins separated in time. Some thought the children's individuality might be threatened, while others argued that identical twins manage to keep their individuality intact.

John Robertson of the University of Texas raises several other issues worth pondering: What about the cloned child's sense of free will and parental expectations? Since the parents chose to duplicate their first child, will the clone feel obliged to follow in the older sibling's footsteps? Will the older child feel he has been duplicated because he was inadequate or because he is special? Will the two have a unique form of sibling rivalry, or a special bond? These are, of course, just special versions of questions that come up whenever a new child is introduced into a family.

Could a megalomaniac decide to achieve immortality by cloning an "heir"?

Sure, and there are other situations where adults might be tempted to clone themselves. For example, a couple in which the man is infertile might opt to clone one of them rather than introduce an outsider's sperm. Or a single woman might choose to clone herself rather than involve a man in any way. In both cases, however, you would have adults raising children who are also their twins—a situation ethically indistinguishable from the megalomaniac cloning himself. On adult cloning, ethicists are more united in their discomfort. In fact, the same commission that was divided on the issue of twins was unanimous in its conclusion that cloning an adult's twin is "bizarre … narcissistic and ethically impoverished." What's more, the commission argued that the phenomenon would jeopardize our very sense of who's who in the world, especially in the family.

How would a human clone refer to the donor of its DNA?

"Mom" is not right, because the woman or women who supplied the egg and the womb would more appropriately be called Mother. "Dad" isn't right, either. A traditional father supplies only half the DNA in an offspring. Judith Martin, etiquette's "Miss Manners," suggests, "Most honored sir or madame." Why? "One should always respect one's ancestors," she says, "regardless of what they did to bring one into the world."

That still leaves some linguistic confusion. Michael Agnes, editorial director of Webster's New World Dictionary, says that "clonee" may sound like a good term, but it's too ambiguous. Instead, he prefers "original" and "copy." And above all else, advises Agnes, "Don't use 'Xerox.'"

A scientist joked last week that cloning could make men superfluous. Is it true?

Yes, theoretically. A woman who wanted to clone herself would not need a man. Besides her DNA, all she would require are an egg and a womb—her own or another woman's. A man who wanted to clone himself, on the other hand, would need to buy the egg and rent the womb—or find a very generous woman.

What are the other implications of cloning for society?

The gravest concern about the misuse of genetics isn't related to cloning directly, but to genetic engineering—the deliberate manipulation of genes to enhance human talents and create human beings according to certain specifications. But some ethicists also are concerned about the creation of a new (and stigmatized) social class: "the clones." Albert Jonsen of the University of Washington believes the confrontation could be comparable to what occurred in the 16th century, when Europeans were perplexed by the unfamiliar inhabitants of the New World and endlessly debated their status as humans.

Whose pockets will cloning enrich in the near future?

Not Ian Wilmut's. He's a government employee and owns no stock in PPL Therapeutics, the British company that holds the rights to the cloning technology. On the other hand, PPL stands to make a lot of money. Also likely to cash in are pharmaceutical and agricultural companies and maybe even farmers. The biotech company Genzyme has already bred goats that are genetically engineered to give milk laced with valuable drugs. Wilmut and other scientists say it would be much easier to produce such animals with cloning than with today's methods. Stock breeders could clone champion dairy cows or the meatiest pigs.

Could cloning be criminally misused?

If the technology to clone humans existed today, it would be almost impossible to prevent someone from cloning you without your knowledge or permission, says Philip Bereano, professor of technology and public policy at the University of Washington. Everyone gives off cells all the time—whenever we give a blood sample, for example, or visit the dentist—and those cells all contain one's full complement of DNA. What would be the goal of such "drive-by" cloning? Well, what if a woman were obsessed with having the child of an apathetic man? Or think of the commercial value of a dynasty-building athletic pedigree or a heavenly singing voice. Even though experience almost certainly shapes these talents as much as genetic gifts, the unscrupulous would be unlikely to be deterred.

Is organized religion opposed to cloning?

Many of the ethical issues being raised about cloning are based in theology. Concern for preserving human dignity and individual freedom, for example, is deeply rooted in religious and biblical principles. But until last week there had been surprisingly little theological discourse on the implications of cloning *per se.* The response so far from the religious community, while overwhelmingly negative, has been far from monolithic.

Roman Catholic, Protestant, and Jewish theologians all caution against applying the new technology to humans, but for varying reasons. Catholic opposition stems largely from the church's belief that "natural moral law" prohibits most kinds of tampering with human reproduction. A 1987 Vatican document, *Donum Vitae*, condemned cloning because it violates "the dignity both of human procreation and of the conjugal union."

Protestant theology, on the other hand, emphasizes the view that nature is "fallen" and subject to improvement. "Just because something occurs naturally doesn't mean it's automatically good," explains Max Stackhouse of Princeton Theological Seminary. But while they tend to support using technology to fix flaws in nature, Protestant theologians say cloning of humans crosses the line. It places too much power in the hands of sinful humans, who, says philosophy Professor David Fletcher of Wheaton College in Wheaton, Illinois, are subject to committing "horrific abuses."

Judaism also tends to favor using technology to improve on nature's shortcomings, says Rabbi Richard Address of the Union of American Hebrew Congregations. But cloning humans, he says, "is an area where we cannot go. It violates the mystery of what it means to be human."

Doesn't cloning encroach on the Judeo-Christian view of God as the creator of life? Would a clone be considered a creature of God or of science?

Many theologians worry about this. Cloning, at first glance, seems to be a usurpation of God's role as creator of humans "in his own image." The scientist, rather than God or chance, determines the outcome. "Like Adam and Eve, we want to be like God, to be in control," says philosophy Prof. Kevin Wildes of Georgetown University. "The question is, what are the limits?".

But some theologians argue that cloning is not the same as creating life from scratch. The ingredients used are alive or contain the elements of life, says Fletcher of Wheaton College. It is still only God, he says, who creates life.

Would a cloned person have its own soul?

Most theologians agree with scientists that a human clone and its DNA donor would be separate and distinct persons. That means each would have his or her own body, mind, and soul.

Would cloning upset religious views about death, immortality, and even resurrection?

Not really. Cloned or not, we all die. The clone that outlives its "parent"—or that is generated from the DNA of a dead person, if that were possible—would be a different person. It would not be a reincarnation or a resurrected version of the deceased. Cloning could be said to provide immortality, theologians say, only in the sense that, as in normal reproduction, one might be said to "live on" in the genetic traits passed to one's progeny.

2

Our Genes Are Not Us

Robert Wright

Your clone might be eerily like you. Or perhaps eerily like someone else.

The world has had a week to conjure up nightmare scenarios, yet no one has articulated the most frightening peril posed by human cloning: rampant self-satisfaction. Just consider. If cloning becomes an option, what kind of people will use it? Exactly—people who think the world could use more of them; people so chipper that they have no qualms about bestowing their inner life on a dozen members of the next generation; people, in short, with high self-esteem. The rest of us will sit there racked with doubt, worried about inflicting our tortured psyches on the innocent unborn, while all around us shiny, happy people proliferate like rabbits. Or sheep, or whatever.

Of course, this assumes that psyches get copied along with genes. That seems to be the prevailing assumption. People nod politely to the obligatory reminder about the power of environment in shaping character. But many then proceed to talk excitedly about cloning as if it amounts to Xeroxing your soul.

What makes the belief in genetic identity so stubborn? In part a natural confusion over headlines. There are zillions of them about how genes shape behavior, but the underlying stories spring from two different sciences. The first, behavioral genetics, studies genetic differences among people. (Do you have the thrill-seeking gene? You do? Mind if I drive?) Behavioral genetics has demonstrated that genes matter. But does that mean that genes are destiny, that your clone is you?

Enter the second science, evolutionary psychology. It dwells less on genetic difference than on commonality. In this view, the world is already chock-full of virtual clones. My next-door neighbor—or the average male anywhere on the globe—is a 99.9 percent-accurate genetic copy of me. And paradoxically, many of

---

This appeared as "Can Souls Be Xeroxed?" *Time*. v149 (March 10 1997). p. 73.

the genes we share empower the environment to shape behavior and thus make us different from one another. Natural selection has preserved these "malleability genes" because they adroitly tailor character to circumstance.

Thus, though some men are more genetically prone to seek thrills than others, men in general take fewer risks if married with children than if unattached. Though some people may be genetically prone to high self-esteem, everyone's self-esteem depends heavily on social feedback. Genes even mold personality to our place in the family environment, according to Frank Sulloway, author of *Born to Rebel*, the much discussed book on birth order. Parents who clone their obedient oldest child may be dismayed to find that the resulting twin, now lower in the family hierarchy, grows up to be Che Guevara.

This malleability could, in a roundabout way, produce clones who are indeed soul mates. Your clone would, after all, look like you. And certain kinds of faces and physiques lead to certain kinds of experiences that exert certain effects on the mind. Early in this century, a fledgling effort at behavioral genetics divided people into such classes as mesomorphs—physically robust, psychologically assertive—and ectomorphs—skinny, nervous, shy. But even if these generalizations hold some water, it needn't mean that ectomorphs have genes for shyness. It may just mean that skinny people get pushed around on the junior-high playground and their personality adapts. (This is one problem with those identical-twins-reared-apart studies by behavioral geneticists: Do the twins' characters correlate because of "character genes" or sometimes just because appearance shapes experience which shapes character?).

People who assume that genes are us seem to think that if you reared your clone, you would experience a kind of mind meld—not quite a fusion of souls, maybe, but an uncanny empathy with your budding carbon copy. And certainly empathy would at times be intense. You might know exactly how nervous your frail, gawky clone felt before the high school prom or exactly how eager your attractive, athletic clone felt.

On the other hand, if you really tried, you could similarly empathize with people who weren't your clone. We've all felt an adolescent's nervousness, and we've all felt youthfully eager, because these feelings are part of the generic human mind, grounded in the genes that define our species. It's just that we don't effortlessly transmute this common experience into empathy except in special cases—with offspring or siblings or close friends. And presumably with clones.

But the cause of this clonal empathy wouldn't be that your inner life was exactly like your clone's (it wouldn't be). The catalyst, rather, would be seeing that familiar face—the one in your high school yearbook, except with a better haircut. It would remind you that you and your clone were essentially the same, driven by the same hopes and fears. You might even feel you shared the same soul. And in a sense, this would be true. Then again, in a sense, you share the same soul with everyone.

3

Clones Are Not Exact Copies

Marc Zabludoff

In February, at the annual meeting of the American Association for the Advancement of Science, in Philadelphia, I attended a session entitled "The Rights and Wrongs of Cloning Humans." Among the speakers was Ian Wilmut, of the Roslin Institute in Edinburgh. Now, Ian Wilmut, as the leader of the research team that cloned the famous lamb Dolly, is rightly the "father" of mammalian cloning, if such a title can be conferred. That is to say, Dr. Wilmut is a man who has unblinkingly beheld the concrete future, has seen it take shape under his own guidance. He is a balding, bespectacled, soft-spoken scientist, with a Scottish accent characteristically clipped but languorous. For all his world-shattering achievement, he is an undramatic and unemotional speaker. Yet there was no mistaking the passion of his sentiments on this subject. When faced with the prospect of colleagues racing to bring forth a cloned human infant, he was appalled.

First of all, he noted, the limits of current technology simply do not allow the attempt: to get one successful birth, many babies would have to die in failed procedures—an absolutely unacceptable price. But even assuming we solve the technical problems, why, he asked, would we want to clone ourselves? Even if we truly desire an exact duplicate of someone—ourselves, a lost loved one, a scientific or artistic genius—the plain truth is that we won't get it.

We are more than our genes. We are our genes in a particular place and time, whole people interacting with others in an infinitely variable world. Only through that experience do we become who we are. A cloned Einstein reared in twenty-first-century Los Angeles will not become a tousled professor of new physics. A cloned Mozart will not re-elevate our souls or drive a cloned Salieri to distraction. A clone of a child tragically and prematurely dead will not replace wholly and without distinction

---

This article appeared as "Fear and Longing" in *Discover*. v19, 1998.

the child who once was. All the clone will be for certain is the bearer of unmet expectation.

That cloning won't fully work should be evident to all of us. We are each a half-clone, after all, with respect to either parent. And though we may at some time have heard that we got, say, our singing voice from our mother or our temper from our father, we know it's not strictly true. Talents and temperament aren't really divvied up, trait by trait, and served intact down the genetic line. Not one of us is identical to a parent, not even in the middle of an aria or rage. And we would not be so even if we were the inheritor of all a parent's genes rather than half.

To be fair, cloning is not the focus of most biotech research. It has simply garnered the most publicity. But it does most dramatically illustrate what some have called the technological imperative—which means that if we can do something, we will, whether there is wisdom in the enterprise or not.

And cloning is not the only application of biotechnology that even the science-supporting public—or the staff of *Discover*—finds disquieting. Yes, we're eager to have the entire human genome laid out before us. Yes, we're eager to see the day when the genes that cause truly terrible diseases can be repaired. But are we ready for casual tinkering with the genes of plants that feed us? Are we ready for the genetic manipulation of cows and pigs for the sole purpose of human convenience? Are we ready to turn loose all the forces of technology to further—surely unnecessarily—the pace of human reproduction, no matter what the material, societal, and psychological costs?

We are preparing for our children a new world, and we don't yet know its borders.

4

Artificial Twinning Is Not Cloning

Josie Glausiusz

In 1979 a Danish scientist named Steen Willadsen took a two-celled sheep embryo, cut open its zona pellucida—the protective, jellylike covering that holds the cells together during the earliest stages of development—and gently eased the two cells apart. Enveloping each cell in an artificial zona of agar (a gelatinous seaweed extract), he then transferred them into the oviduct of a ewe. There each cell continued to divide on its own. What Willadsen had shown was that each cell could give rise to a separate embryo, and ultimately to a lamb. His work on sheep and later on cows has proved useful to cattle breeders: it allows them to produce identical offspring of prize animals by splitting a single fertilized embryo.

Last October, Jerry Hall and Robert Stillman, *in vitro* fertilization specialists at George Washington University Medical Center in Washington, D.C., announced that they had taken a first tentative step toward repeating Willadsen's experiment—but this time with human embryos. In describing their work at the annual meeting of the American Fertility Society, Hall and Stillman said they had "cloned" human embryos. The use of that word, with its chilling overtones helped ignite the debate that was soon joined on the nation's front and op-ed pages by religious gious leaders and medical ethicists.

What did Hall and Stillman actually do? First of all, it was not cloning in the technical sense. True cloning implies reproduction without sex. It would mean creating an exact copy of an adult human—by taking a single cell from that person, placing it inside a human egg cell that has had own genes and indeed its entire nucleus removed, and allowing that single cell to grow into a new adult as a normal embryo would. This has never been done in mammals, let alone humans, because each cell of an adult

---

This appeared as "Splitting Heirs" in *Discover*. v15 (January 1994). pp. 84–5.

mammal has a specific role—it's a brain cell, say, or a muscle cell—and no longer has the ability to divide and grow into all the other kinds of cells that are needed to make a whole animal. Human cells lose this "totipotency" when the embryo consists of just eight cells.

Hall and Stillman experimented on totipotent cells from 17 human embryos. Eight consisted of just two cells; the rest were already at either the four- or eight-cell stage. The researchers dissolved the zona pellucida with enzymes and separated the individual cells, which are called blastomeres. They then placed each blastomere in a seaweedextract gel, just as Willadsen had done with his sheep cells. Unlike Willadsen, they never implanted their cells in a mother—they just watched as each cell, isolated from its neighbors, began to divide anew in a laboratory culture.

Not surprisingly, cells from the youngest embryos did best. Blastomeres from the four-cell stage divided only four times, and ones from the eight-cell stage only three times, before they stopped, apparently because they had run out of cytoplasm—the intracellular goo they need for making new cells. (At this stage cells don't grow between divisions, so their cytoplasm keeps halving.) But the cells from two-cell embryos started out larger, and so they were able to divide five times, reaching the 32-cell stage. That's the earliest stage at which an embryo created through *in vitro* fertilization would normally be implanted in the mother's uterus.

And that's the main technical purpose of Hall and Stillman's research: to add another tool to the kit of *in vitro* fertilization practitioners. It is already common practice to implant three or more individually fertilized embryos in a prospective mother's uterus to increase the likelihood that at least one will "take." But women undergoing IVF treatment must be given fertility drugs to make them produce large numbers of eggs, and not all women succeed even then. Embryo splitting *à la* Hall and Stillman might offer a solution to this problem—with the difference that the implanted embryos would be genetically identical twins. Although the work was hardly a major technical breakthrough, the American Fertility Society was impressed enough to award Hall and Stillman a prize for their presentation.

But the researchers also claim to have had another purpose in splitting human embryos. "It was clear that it was just a matter of time before someone was going to do it," Hall told the journal Science before he stopped talking to the press, "and we decided it would be better for us to do it in an open manner and get the ethical discussion moving." In that respect he and Stillman clearly succeeded—although the discussion was led astray to some extent by their description of their work as "cloning" rather than "embryo splitting" or "artificial twinning."

Some embryo researchers responded with outrage to the misnomer, fearing that the specter of mass-produced humans (hordes of Dan Quayles and Ross Perots were among the fearsome possibilities raised in the press) would endanger government funding for their own research. "The public will be horrified, and quite naturally so," says Jacques Cohen, scientific director of the *in vitro* fertilization program at New York Hospital–Cornell University Medical College. Not only was what Hall and Stillman did not cloning, says Cohen, it was "a silly little experiment."

Silly or not, cloning or not, the experiment clearly raises ethical issues, and Hall and Stillman are waiting for the dust to settle before continuing their research. Could a couple freeze an embryonic copy of their *in vitro*-fertilized child, then raise the identical twin years later, once they've seen how the first one turned out? Could they "reserve" a disposable embryo for genetic testing or even as an organ donor for the "keeper"? How about selling an embryo alongside a glowing resume of the child it will turn out to be? "Things like that are to me a little outlandish," says Robert Visscher, executive director of the American Fertility Society. "From our perspective, we're always interested in basic research that gives rise to technology that will improve the chances of an infertile couple's achieving their goal of having children. But you can't totally predict how technology will be used in the future, and it's for society to decide what's the appropriate use and what isn't."

5

There's Dispute over the Moral Status of a Cloned Cell

Vincent Kiernan

Almost a year after Scottish researchers announced that they had cloned an adult sheep, the issue of whether it is proper for scientists to clone human cells for research and therapeutic purposes has become enmeshed in one of the most contentious issues in American society: When does human life begin?

Critics of the technology say cloning a human cell would generate an embryo that has the full moral status of any other human and therefore should not be experimented upon or treated in any way other than being allowed to develop and be born. Supporters of the technology say cloning produces a set of cells that may or may not develop into an embryo, depending on the conditions under which they are kept, and so should not be seen as morally equivalent to humans.

The contrasting views were in evidence in comments by researchers, ethicists and lawmakers at the annual meeting of the American Association for the Advancement of Science, which ended here last week, and in Washington at a hearing of the House Commerce Subcommittee on Health and Environment, which was discussing the possibility of banning the cloning of human cells.

When a Clone Acquires Rights is an Issue

"Although it is a potential human, I don't think in many important aspects it really is a human," Ian Wilmut of the Roslin Institute, a biomedical-research facility in Scotland, told the science conference. He headed the team that produced the cloned sheep, Dolly.

This appeared as "Debate Over Cloning Touches One of Society's Most Sensitive Nerves" *Chronicle of Higher Education*, v44 n25, pp. A16–A17, Feb 27, 1998.

But just the day before, Dianne N. Irving, a biochemist and professor of the history of philosophy and medical ethics at the Dominican House of Studies, a Roman Catholic seminary in Washington, told the House subcommittee that cloning a human cell would produce a new human life.

"This human being, who is a single-cell human embryo, or zygote, is not a 'potential' or 'possible' human being, but is an already existing human being—with the 'potential' or 'possibility' to simply grow and develop bigger and bigger," she said.

The question of the moral status of a cloned cell is not merely academic. It is expected to play a key role in whatever decision Congress reaches about banning or restricting the use of cloning with human cells.

Several bills to ban or restrict cloning are pending. Two weeks ago, the Senate voted not to consider a bill that would have barred all use of Dr. Wilmut's technique with human cells (*The Chronicle,* February 20).

A rival Senate bill would allow research with cloning but not the implantation of a cloned embryo into a human uterus. Another bill, approved by the House Science Committee in August, would bar the use of federal funds in human cloning research.

"They want to continue research with embryos, and we want to stop it," said Representative Vernon J. Ehlers, a Michigan Republican who sponsored the cloning bill pending in the House. "What it really comes down to is: What rights does the clone have at any stage of its existence?" said Mr. Ehlers, a physicist.

Classifying a cloned human cell as a human embryo also would have the consequence of strangling university research into therapeutic and research uses of cloning, because faculty members would wish to avoid the controversy surrounding such a highly charged topic, said Michael D. West.

"A lot of researchers wouldn't go into the field for fear their reputations would be damaged," Dr. West, founder of two biotechnology companies and spokesman for the Biotechnology Industry Organization, told lawmakers at the Congressional hearing. "It would be like setting a minefield in front of the university."

Whether a Cloned Cell is an Embryo is a Bone of Contention

Animals and plants are composed of many types of cells,

with specialized functions—nerve cells behave differently from muscle cells, for example. However, every cell contains all of the genetic programming needed for any type of cell. Any genetic information that is not pertinent to that cell's particular function is deactivated; for example, muscle cells contain a deactivated version of the DNA that tells nerve cells how to behave.

Dr. Wilmut's experiment to produce Dolly managed, for the first time, to reactivate dormant genetic material in an adult cell. In its technique, known as somatic cell nuclear transfer, his team removed the genetic material from a sheep egg and placed it next to a cell from another sheep's udder. When the researchers applied an electrical shock to the mammary cell, the two cells fused. The egg reactivated dormant genetic material that had come from the mammary cell, and the egg developed into a sheep fetus.

As the cloned cell divides, it develops first into a mass of "totipotent" cells, which have the ability to develop into any kind of sheep cell, not just a mammary cell like the one that was cloned. From this collection of cells, a sheep embryo develops.

Dr. Wilmut and others believe that, if the totipotent cells are grown under the right conditions, they could be coaxed to develop into a mass of specific tissues. For example, they might produce heart-muscle cells, which could be used to repair cardiac damage, or skin cells, which could be used to treat burns.

And if those tissues were produced by cloning cells from the patient's own body, the patient would not reject the transplanted tissues, because they would be genetically identical to his or her own.

Religious Leaders Say It's Immoral to Create Human Life in Order to End It

Dr. Wilmut said he believed that such medical uses of cloned cells are ethical. "It is something that I would consider doing," he told the science conference.

But William Cardinal Keeler of Baltimore, representing the Roman Catholic bishops of the United States, told lawmakers that the procedure would be immoral, because it "would require creating, developing, and then killing a human embryo that is the patient's identical twin."

"Creating human life solely to cannibalize and destroy it is the most unconscionable use of human cloning, not its highest justification," he said.

A similar sentiment was expressed by Rabbi Barry Freundel,

of Kesher Israel Congregation in Washington, who told lawmakers that it would be immoral to create embryos that would be used in the process of giving medical treatment to another person.

The question of the morality of the use of cloning in research has been ignored by most Christian denominations, Ronald S. Cole-Turner, a professor of theology and ethics at Pittsburgh Theological Seminary, told the science conference. Instead, he said, religious leaders have focused their attention on the issue of the use of cloning to produce humans.

One exception, he said, has been the United Church of Christ, which "left the door open to human-embryo research" with a statement issued last year that would countenance research on cloned "pre-embryos" for 14 days after their creation, as long as the research is "well-justified" and shows "proper respect for the pre-embryos."

Some Deny that Cloned Embryos Are Potential Persons

Some say that because it is not inevitable that a cloned cell will develop into an embryo, the cloned cell need not be regarded as human. Arthur L. Caplan, director of the Center of Bioethics at the University of Pennsylvania, noted that the embryo produced by the cloning process needed careful handling and treatment to grow into a fetus.

"I'm not sure that I'm ready to say that cloned embryos are potential persons," he told reporters at the science conference. Moreover, he said, if cloned cells are human, then the line of argument could be extended to confer moral status on cells such as ordinary skin cells—that could be used to produce the cloned embryos.

By that line of thinking, "any one of our cells could be a potential person," Dr. Caplan said.

At the House hearing, lawmakers who want to avoid restrictions on therapeutic or research uses of cloning sounded a similar theme. "Somatic-cell nuclear transfer is just the latest tool of genetic research. It is not the same as cloning a human being," said Representative Henry A. Waxman, a California Democrat.

Representative Greg Ganske, an Iowa Republican and a surgeon, said cloned skin cells could be important in treating patients with severe, extensive burns. He conceded that a cloned cell might divide and grow under laboratory conditions, but said: "That doesn't mean they would have to develop into a fetus. The addition of growth factors could insure that the cells develop into specialized tissues, not a person."

"It's not clear that it would be the creation of an embryo," said Dr. West, of the biotechnology organization. He said the cloned cell should be thought of as a "primordial stem cell," an unspecialized cell that can develop into any kind of cell. In laboratory experiments with mice, primordial stem cells do not always produce a fetus when implanted in the uterus, he said.

Future Cloning Techniques Might Get Round Religious Objections

Others disagree. Primordial stem cells are simply human embryos by another name, the Rev. Kevin T. Fitzgerald, a research associate in molecular genetics at the Loyola University Medical Center in Chicago, told the House subcommittee. "A rose is a rose is a rose."

"If an embryo is dividing and developing, it is a member of the human family and deserves our respect," Representative Dick Armey, a Texas Republican and House majority leader, said in a statement released at the hearing.

The issue of the moral status of the cloned cells could be skirted if scientists could find a way to produce only medically valuable cells through cloning, without passing through the intermediary step of producing a set of cells that could develop into a fetus.

Indeed, Cardinal Keeler told lawmakers that the Catholic Church would raise no moral objections to cloning cells in this fashion.

"If you don't have a new human being, we have no problem," he said.

But such a prospect seems far off, because researchers have only vague notions of how to manipulate a cloned cell so that it does not exist, even briefly, as an embryo. "Biologically, we have a lot to learn," said Dr. Wilmut.

"Who knows how long 'long term' is in this context?" he added. But when it happens, the development will be another major advance in the field, he said. "That is going to be, if you like, the next Dolly experiment."

6

We're a Long Way from Designer Babies

Caroline Daniel

Genome? Is that a manipulated gnome?

The genome is a compendium of all genetic instructions. Written along the double helix of DNA are three billion or so chemical letters of genetic text, contained in more than 100,000 genes. The project aims to identify every single letter. This is the easy bit. The hard bit will be to work out how all the genes work, how they interact and how they respond to diseases and to the environment.

Whose project is it?

It is an international scientific collaboration, set up in 1990 and largely public-sector funded to the tune of £2 billion-plus. The main backers are American: the National Institutes of Health and the U.S. Department of Energy. Other genomes, such as mice and yeast, are also being mapped, so we'll be able to work out how far man is, in fact, mouse, or even risen mouse.

Will this public-sector project ever be finished?

The target is to sequence all the genes in the human genome by 2005. Unlike the Jubilee Line, it's on schedule. As of late 1997, 50,000 genes had been mapped.

What's the point of it?

Gene therapy. The original thinking was that we could cure genetic disorders (4,000 conditions are linked to single genes). If you could identify the mutant gene, it was hoped, you could replace it with a working copy and, hey presto: a cure.

This appeared as "Human Genetics: The Dinner-Party Guide" in *New Statesman*, v41 (13 February 1998), p. 31.

Hang on, you said "hoped". What's gone wrong?

The first experiment for gene therapy was in 1991 for adenoise deaminase (better known as "boy-in-the-bubble disease", except in this case it was a girl). The girl now leads a normal life. But because she was given lots of treatments at the same time, we can't say for sure that the gene therapy worked. Other trials have not been a huge success. In fact, in 1995 the NIH declared that less cash should go on clinical trials and more on basic science. About 250 gene-therapy trials are under way worldwide.

Why so little success?

The problem is how to get the new genes to the right places. One way is to send a thief to catch a thief, by using a virus. Because viruses naturally infect certain cells, someone had the bright idea of inserting therapeutic genes into them, so they could hitch a ride to the cells. Unfortunately, the immune system can react badly to the virus.

Does that mean there aren't any medical benefits? No. The latest technique, invented by a former NIH scientist, Craig Venter, decodes the messages sent out by genes in the form of proteins. From this we will be able to develop therapeutic proteins that counteract the effects of dodgy genes or even help to boost the immune system against diseases.

When will that be?

It's happening. In December the U.S. firm Human Genome Sciences announced that the first drug developed from the Genome Project was about to be tested on humans. It could prevent damage caused to bone marrow by chemotherapy. And don't forget, once a gene has been identified you can test for it. For now, testing is confined to serious single-gene disorders, such as some breast cancers. But these account for about 2 percent of human maladies; in most diseases the contribution of genes is fuzzier. An environmental trigger may also be required, or several genes need to be faulty. So we are a long way away from screening for most diseases.

And a long way from being able to screen out hereditary diseases before birth?

Not necessarily. Doctors use some of these tests to do preimplantation diagnoses on embryos. They can then select an embryo which is free of a nasty gene, such as the one that leads to cystic fibrosis. If it were possible to test for an eye colour gene, or baldness gene, you could theoretically screen for these. In practice the authorities, at least in this country, wouldn't let you.

So no designer babies, then.

In America last year newspapers carried full-page ads: "What if you could choose exactly what your child was like? Male or female? Gay or straight? You decide," said the headline, next to a photo of a cute-looking baby. Underneath was a shopping list of traits you could pick, such as skin colour or musical ability, and a phone number to call ...

Designer babies!

Relax. It's just the blurb for a new film called *Gattaca*, about a society run by a genetically engineered elite. With today's technology we cannot design. To do that, you have to be able to target all those cute designer genes at the embryo stage.

But the targeting will get easier ...

Even if you can tweak your embryo you still have to implant it in a womb. Fewer than one in six embryos transferred using IVF leads to a successful pregnancy, a rate that is improving only very slowly. And if we're some way from being able to pick eye colour, we're further off eliminating temper tantrums or crafting mini-Einsteins. The scientists have found it very hard to identify single genes for complex forms of behaviour. Announcements such as "gene for risk-taking found" may kindle the hopes of bridge fans eager to screen future partners, but are seriously premature.

So my family's chronic-untidiness gene is with us for a few generations yet.

Once we have the entire genome mapped, it will be easier to assess the role of genes in behaviour. We will be able to do psychological and genetic profiling by crunching the numbers to see what genetic coincidences there are. That's why some fear a redefinition of individuals in terms of their DNA. But the rising tide of genetic information is bound over time to alter our perceptions of what humanity is like.

II

First Reactions to Dolly

Introduction to Part II

The cloning of Dolly by Ian Wilmut was a surprise and a shock. In Part II we focus on the kinds of responses it provoked. Reading 7, by Bob Harris, is an irreverant, tongue-in-cheek look at some questions raised by cloning, with a serious undertone. Beth Baker (Reading 8) describes the frantic ripples of the Dolly innovation among leading American politicians. The Editors of *National Review* (Reading 9) express the immediate hostility to the prospect of human cloning voiced by many Christian conservatives. However, in Reading 10 Ruth Macklin offers the opposing point of view that, while cloning should be subject to legal control, there is not yet any convincing reason to prohibit cloning.

Christopher Wills gives a more detailed argument, in Reading 11, that some of the strong antipathy to human cloning may be hard to defend rationally. In Reading 12, Virginia Morell provides some fascinating factual information, while evidently sympathetic to the view that cloning has something to offer childless couples.

7

Cloning Forces Us to Face New Puzzles

Bob Harris

Scientists announced last February that they had cloned the first large adult mammal. Despite President Clinton's immediate response that no federal funds will be used for research on human cloning (because, in his words, "human life is unique, born of a miracle," a "profound gift," and all that), human cloning is surely inevitable. This means we're forced to face some new and serious questions.

For example, since cloning will be expensive initially, won't only rich people be cloned? Will Donald Trump be cloned in the womb by hundreds of women who need the money? Is it possible that a century from now there will be entire cities of Perots and Forbeses?

If not, won't clones still be considered status symbols, displayed at cocktail parties and on the cover of *InStyle* magazine? Or will the replication of the rich simply dilute their wealth?

Will a black market arise, from which the poor can get back-alley clones?

Will actually being a clone thereafter carry a certain élan? Or will it be more déclassé, like owning a print of an oil painting instead of the original?

Once cloning can begin *in utero*, how will we tell clones from originals? Dental records? Tattoos? Certificates of authenticity that are carried along with drivers' licenses?

Will clones be subconsciously considered disposable? Will killing a clone carry less of a stigma than murdering the original?

How long until some rich individual creates lobotomized "spares" to replace his or her own aging human body parts? Will wealthy parents hire surrogates and have their children in batches of four or five so there are "extras" if one gets hit by a car?

This appeared as "Second Thoughts About Cloning Humans" in *The Humanist*, v57 (May/June 1997). p. 43.

What social security numbers do clones get? Do we just add a letter to the donor's number, starting with A for the first clone, B for the second, and so on?

Since most replicants will be born into wealth, will we see outbreaks of envious blue-collar clone-bashing? Will clones, like other oppressed groups, develop a system of nonverbal behaviors—such as wearing lapel pins shaped like rubber stamps—to signify their status? Will clones develop a national support network (ACNE: Adult Children of Nobody, Exactly)?

In school, will clones be allowed to copy on exams?

Since only a small percentage of cloned zygotes survive the process, how long until pro-lifers begin bombing chemistry labs?

Since DNA can be furtively collected from such things as used facial tissues, how long until someone is cloned against her or his will? Will it be a crime? With what punishment? Who gets custody of the clone?

Will professional sports have strict anti-cloning rules? If not, how much will Michael Jordan's toenail clippings be worth?

Will cloning a second set of kids become a custody option in divorce proceedings? If someone who has been cloned dies without a will, who gets the stuff? The family or the clone?

If a clone has *déjà* vu, how can he or she tell?

If one accepts the Catholic notion of new soul at conception, when exactly does a clone's soul form? If without conception there's no new soul, does the clone timeshare with the original? What happens if the donor is saved and the replicant isn't? If clones have no soul, can they sing soul music convincingly?

Since DNA can be recovered from the dead, what's the status of the clone's soul then? Since clones can be born to a virgin mother, would they therefore be holy?

How long until someone attempts to validate the Shroud of Turin by scraping off some DNA, raising the kid, and seeing if he can transform water into wine? If a Jesus clone goes to church, will he sit in the audience or onstage? When the clone starts advocating humility, pacifism, and aid to the needy, how long until he is crucified?

Since DNA evidence will become all but meaningless, what will attorney Barry Scheck do for a living?

If a woman has a *ménage à trois* with her husband and his clone, has she violated her wedding vows? Is a child sired by the clone illegitimate?

Masturbation isn't generally considered a crime. How about touching your clone in a sexually arousing way?

And how long until cloning is outlawed by male-dominated legislatures—just as soon as the men realize that women no longer need them?

8

Politicians Were Agitated by Dolly

Beth Baker

"Sorry I didn't return your call sooner," a harried Congressional press aide told *BioScience*. "But we've been overrun by sheep."

So it went on Capitol Hill following the 22 February announcement that a Scottish research team had successfully cloned a sheep from a six-year-old ewe. The humble sheep, Dolly, became an overnight media sensation and had members of Congress scurrying to react.

The chief concern was whether sheep cloning would soon lead to human cloning. This fear was heightened when the Oregon Regional Primate Research Center revealed just a few days later that it had cloned rhesus monkeys using nuclear transfer technology. Although this is a well-established method involving the transfer of a full set of chromosomes from one embryo cell to another, the process had not been used successfully in primates until now.

Representative Connie Morella (Republican–Maryland), chair of the House Technology Subcommittee, was first out of the block with a packed 5 March hearing. "The result of these successful experiments with sheep and monkeys, with the potential for the future extension to humans, is ... provoking worldwide discussions on the ethics and morality of cloning," Morella said.

Morella's hearing was topped a week later when Senator Bill Frist (Republican–Tennessee), chair of the Senate Subcommittee on Public Health and Safety, held his own media-filled hearing, this one featuring the head of the Scottish research team, Ian Wilmut.

This appeared as "To Clone or Not To Clone—Congress Poses the Question" in *BioScience*, v47 (June 1997). p. 340.

Frist, a former heart transplant surgeon, stressed the importance of getting Congress and the public to understand better the science and the potential benefits of cloning. When heart transplants were new, they elicited similar alarms, he recalled. "The science was vilified," he said.

At both hearings, questions from legislators to the panels of scientists and ethicists ranged from the mundane ("Having considerable experience with protecting a small fly in my district, I was wondering, could we clone it and move it somewhere else?") to the metaphysical ("How will cloning affect how we view the spirit or soul?").

A number of legislators inquired hopefully if cloning meant we need not worry about endangered species anymore, but the assembled scientists seemed reluctant to encourage this notion. "Cloning might not be the best avenue for preservation of species," said a diplomatic Harold Varmus, director of the National Institutes of Health.

Wilmut predicted that within two to three years, cloned farm animals would be producing proteins in their milk that would be used to treat human diseases such as hemophilia. This breakthrough would soon be followed by using cloned animals to study cystic fibrosis and other genetic diseases. Susan Smith, director of the Oregon Regional Primate Research Center, said her team's research with cloned rhesus monkeys would help sort out the role of genes and the environment in human disease.

But legislators continued to bring the discussion back to their worries over human cloning. The assembled scientists were unanimous in their belief that cloning of adult humans was not a legitimate avenue of research. "(Human cloning) is offensive and not scientifically necessary," said Varmus, who spoke at both hearings.

Wilmut agreed. "I've not heard of an application that I'm comfortable with, as far as copying a person," he said.

But despite these assurances, several members of Congress indicated that there would be pressure on them to act. "Our ability to restrain the outcry depends on public understanding," said Senator Christopher Dodd (Democrat–Connecticut). "How should we approach legislation on this issue?"

"Legislation and science don't mix very well," said Varmus. "Legislation is difficult to reverse." A ban on human cloning research might be interpreted in ways that would block important benefits, Varmus noted. For example, such a ban, if not carefully worded, might prevent scientists from cloning a human skin cell for the purpose of a bone marrow transplant. He and others urged Congress to hold off on passing legislation until, at the very least, a presidential bioethics commission on human cloning had completed its 90-day review in early June.

Some in Congress were in no mood to wait. Bills to prohibit human cloning appeared immediately in both the House and Senate. "Human cloning is morally reprehensible," said Senator Christopher Bond (Republican–Missouri), author of a bill to ban federal support for human cloning. "I don't believe we need to study this any further."

Other members seemed to want to move more slowly on legislation. "Can you write a bill so narrow that you don't jeopardize the very good research that can come of this?" asked Frist.

That sentiment was echoed on the House side by George E. Brown Jr. (Democrat–California), who compared the strong reaction to cloning with congressional inquiries into recombinant DNA research two decades ago. "It is significant that Congress followed the lead of the scientific community in discussing those issues," he recalled. "It caused us to curb our instincts to go out and regulate everything."

9

There Is No Right to Clone

The Editors of *National Review*

On February 24, when the world was abuzz with the appearance of Dolly the cloned sheep, President Clinton appointed a National Bioethics Advisory Commission, asking it to report within ninety days "with recommendations on possible federal actions" to prevent the "abuse" of cloning human beings. On June 7, the commission reported, proposing a legislative ban on human cloning. The commission's recommendation, while welcome, is accompanied by an unacceptable proviso. It suggests that the prohibition against cloning should expire in three to five years unless a body of experts recommends otherwise. Prohibition of cloning people should be permanent.

The objection is raised that, if cloning is banned here, people of means will go to other countries where it is permitted. Perhaps, but Britain has already banned human cloning, and other countries are considering doing so. In any case, the U.S. should take the lead in securing international agreement to outlaw the procedure. Meanwhile, a national ban must include, as the commission recommends, not only federally funded projects but also private initiatives. Reproductive technology has become a very big business and would undoubtedly explode were people offered the chance to manufacture a "perfect baby," meaning a duplicate of themselves or of celebrities admired for their brawn, brains, or beauty.

The liberal claim that people have a "right" to clone themselves or others is meretricious nonsense. Civilized peoples have long banned incest, polygamy, and other forms of "reproductive freedom." Nor is a ban on scientific technology without precedent. It is, for instance, against the law for individuals to produce weapons of mass destruction. The prospect

---

This appeared as "All the Same" by the Editors of *National Review* in *National Review*, v49 (June 30 1997), pp. 17–18.

of a human cloning industry is every bit as threatening to the future of the human community. Rogue scientists may try to evade such a ban, but the law can provide severe punishments that will discourage them from publicly claiming credit and posing as bold pioneers of scientific "breakthroughs".

An effective ban must also include research aimed at human cloning. In particular, the prohibition against the creation of human embryos for research purposes—which the Commission would mistakenly overturn—must be maintained and strengthened. Such so-called pre-implantation embryos are in fact unimplanted embryos. They are human and they are alive. To create developing human lives simply in order to use and then discard them is morally unconscionable. Have no illusions: Cloning is the manufacture of human beings for the use of other human beings, and its nightmarish implications are not the stuff of science fiction. The technological imperative that says whatever can be done must be done, combined with vast financial interests and a propaganda machine that exploits popular sympathy for people who want a baby of their own design, makes the nightmare almost inevitable—unless we draw the line now.

10

There's No Justification for an Outright Ban on Cloning

Ruth Macklin

Last week's news that scientists had cloned a sheep sent academics and the public into a panic at the prospect that humans might be next. That's an understandable reaction. Cloning is a radical challenge to the most fundamental laws of biology, so it's not unreasonable to be concerned that it might threaten human society and dignity. Yet much of the ethical opposition seems also to grow out of an unthinking dust—a sort of "yuk factor." And that makes it hard for even trained scientists and ethicists to see the matter clearly. While human cloning might not offer great benefits to humanity, no one has yet made a persuasive case that it would do any real harm, either.

Theologians contend that to clone a human would violate human dignity. That would surely be true if a cloned individual were treated as a lesser being, with fewer rights or lower stature. But why suppose that cloned persons wouldn't share the same rights and dignity as the rest of us? A leading lawyer-ethicist has suggested that cloning would violate the "right to genetic identity." Where did he come up with such a right? It makes perfect sense to say that adult persons have a right not to be cloned without their voluntary, informed consent. But if such consent is given, whose "right" to genetic identity would be violated?

Many of the science-fiction scenarios prompted by the prospect of human cloning turn out, upon reflection, to be absurdly improbable. There's the fear, for instance, that parents might clone a child to have "spare parts" in case the original child needs an organ transplant. But parents of identical twins don't

---

This appeared as "Human Cloning? Don't Just Say No" in *U.S. News and World Report*, v122 (March 10 1997), p. 64.

view one child as an organ farm for the other. Why should cloned children's parents be any different?

Cloning Need Not Mean Trying for a Super Race

Another disturbing thought is that cloning will lead to efforts to breed individuals with genetic qualities perceived as exceptional (math geniuses, basketball players). Such ideas are repulsive, not only because of the "yuk factor" but also because of the horrors perpetrated by the Nazis in the name of eugenics. But there's a vast difference between "selective breeding" as practiced by totalitarian regimes (where the urge to propagate certain types of people leads to efforts to eradicate other types) and the immeasurably more benign forms already practiced in democratic societies (where, say, lawyers freely choose to marry other lawyers). Banks stocked with the frozen sperm of geniuses already exist. They haven't created a master race because only a tiny number of women have wanted to impregnate themselves this way. Why think it will be different if human cloning becomes available?

So who will likely take advantage of cloning? Perhaps a grieving couple whose child is dying. This might seem psychologically twisted. But a cloned child born to such dubious parents stands no greater or lesser chance of being loved, or rejected, or warped than a child normally conceived. Infertile couples are also likely to seek out cloning. That such couples have other options (*in vitro* fertilization or adoption) is not an argument for denying them the right to clone. Or consider an example raised by Judge Richard Posner: a couple in which the husband has some tragic genetic defect. Currently, if this couple wants a genetically related child, they have four not altogether pleasant options. They can reproduce naturally and risk passing on the disease to the child. They can go to a sperm bank and take a chance on unknown genes. They can try *in vitro* fertilization and dispose of any afflicted embryo—though that might be objectionable, too. Or they can get a male relative of the father to donate sperm, if such a relative exists. This is one case where even people unnerved by cloning might see it as not the worst option.

Even if human cloning offers no obvious benefits to humanity, why ban it? In a democratic society we don't usually pass laws outlawing something before there is actual or probable evidence of harm. A moratorium on further research into human cloning might make sense, in order to consider calmly the grave questions it raises. If the moratorium is then lifted, human cloning should remain a research activity for an extended period. And if it is ever attempted, it should—and no doubt will—take place only with careful scrutiny and layers of legal oversight. Most important, human cloning should be governed by the same laws that now protect human rights. A world not safe for cloned humans would be a world not safe for the rest of us.

11

Clones Will Be Different

Christopher Wills

Nearly anything goes these days in animal breeding, and in that context what Ian Wilmut of Edinburgh's Roslin Institute accomplished was just the logical next step. Wilmut took the cell nucleus from a Finn Dorset sheep, substituted it for the nucleus of an egg from a Poll Dorset, and implanted it in a ewe of yet a third breed, a Scottish Blackface. The surrogate mother then took over, using maternal magic that has yet to be duplicated in the laboratory, and five months later a lamb called Dolly met the world. Last February, the world met Dolly in an explosion of photographic flashes. Of all the thousands of pictures that came out of Scotland that month, one of the more remarkable was the one that showed Dolly the almost-Finn Dorset standing next to her strikingly different surrogate mother. That, after all, had been the point of using the three different breeds: so that it would be readily apparent that Dolly had not gotten her genes from either the egg donor or the surrogate mother—that she really was the first clone of an adult mammal.

Adult frogs had been cloned before, and so had embryonic mammals. Wilmut himself had cloned embryonic sheep a year earlier, fusing a sheep fetal cell with an enucleated egg cell, one that had its nucleus removed, and then implanting that new egg in a ewe. But it was Dolly who catapulted him onto the world stage. Not only did her genes come from an adult, six-year-old sheep but they came from a dead one—from frozen mammary tissue.

Many scientists were astonished by that technical achievement. It had been assumed that adult cells lost their "totipotency," the ability to give rise to a viable, fertile organism. Adult cells are already specialized, and that process had been thought to involve irreversible changes in their genes. For instance, the DNA may get repacked in such a way that some genes are not available for making proteins. Dolly proved, though,

---

This appeared as "A Sheep in Sheep's Clothing?" in *Discover*, v19 (January 1998). pp. 22–3.

that at least some mammary cells in sheep retain their youthful plasticity—they just need to have it restored somehow.

That's where Wilmut's team made its breakthrough. Previous attempts at cloning from adult cells had failed, they decided, because the cells used were too metabolically active, or in the wrong stage of the cell cycle, or had the wrong set of genes turned on. The Scottish researchers compensated for these possibilities by starving the cells for several days before fusing them with enucleated eggs. In this state of deprivation, the cells' DNA-copying machinery ground to a halt, arresting the cell cycle and forcing the cells into a metabolic torpor that presumably matched the quiescence of an unfertilized egg. Even with this technical sleight of hand, however, Dolly was Wilmut's only success in 277 tries.

ABS Global, a Wisconsin company, has since claimed to have improved on the Scottish procedure with a technique that can rejuvenate more than just mammary cells: it says it has created cow embryos from skin, bladder, and udder cells of an adult cow. The cells were fused with enucleated eggs, as at Roslin. But once the fused cells had begun to divide, a single cell was extracted and inserted into another enucleated egg. Once that embryo began to form, it was implanted in a surrogate mother cow. In late October, ABS reported that the pregnancies—all with normal-appearing embryos—were well under way. Apparently the passage through the first embryo had helped the adult nucleus revert to its totipotent state.

Dolly, too, looks fine now. But she began life with a nucleus from an adult cell—and cells age. Their chromosomes shorten at the tips. Has Dolly already begun to suffer such damage, and will she suddenly begin aging prematurely? Alternatively, was her originating nucleus undamaged? Or—and this would be the most exciting of all—can the milieu of the egg cell somehow reverse genetic damage due to aging? Is the egg a cellular time machine?

Those were some of the scientific questions raised by Dolly's birth, but it was the ethical ones that preoccupied the public. Even before Dolly, the British Parliament had banned human cloning. In June, President Clinton proposed a similar ban. "Banning human cloning reflects our humanity," he said. "It is the right thing to do ... At its worst (this new method) could lead to misguided and malevolent attempts to select certain traits, even to create certain kinds of children—to make our children objects rather than cherished individuals."

Biotechnology companies, and the biomedical community in general, reacted negatively to the proposed ban, even though Clinton emphasized that the cloning of animals would still be allowed. Some would argue that cloning is no way to treat animals, either—that they, too, should not be considered mere objects. But if you accept factory-farming them for food and breeding them for medical experiment, as most of us do at least tacitly, then it becomes hard to cavil at cloning. It is just the logical next step.

And it is one that carries enormous promise. Cloning adult animals means quality control. Once you produce an animal with the traits desired, carbon copies will be easy to make. Imagine entire herds of champion milk producers or paddocks full of Triple Crown winners. More plausibly, imagine flocks of sheep or herds of cows producing milk that carries a critical human protein. That is the goal of Wilmut's research, which is financed in part by PPL Therapeutics, a Scottish pharmaceutical company. In July, Wilmut's group announced that it had produced sheep cloned from fetal cells carrying a human gene—and that gene was turned on only in the sheep's mammary glands. Scottish pastures may soon become pharmaceutical factories.

But of course everybody is agog about the possibility of human cloning. Would it necessarily be wrong? One can easily imagine a scenario in which cloning might be a blessing. Suppose a couple has one treasured child and they are unable to have more. Then the child is suddenly killed in an accident. Should the couple be allowed to clone the child and start over? Many of us would say yes.

A Morally Consistent Ban Is Hard to Formulate

Moreover, practices that our society already accepts as ethical make it harder to stake out a morally consistent ban on cloning. If a woman already has the right to conceive a child with anonymous sperm from a sperm bank, *in vitro* if necessary, with a surrogate mother to bear it if necessary, on what principle do we deny her the right to clone herself? Or a man, if we don't want to be sexist? In humans as in animals, cloning is in some ways just the logical next step. Some would say we've been falling since the top of the stairs, others that we've been climbing to new heights of human liberty—but the point is that once you start on such a path it is hard to stop.

At the moment, technical obstacles to cloning create ethical ones: one reason not to try to clone a human now is that the job would most likely be botched. Wilmut acknowledged this when he testified before a U.S. Senate subcommittee on public health last March. "In previous work with cells from embryos, three out of five lambs died soon after birth and showed developmental abnormalities," he said. "Similar experiments with humans would be totally unacceptable."

Clones Would Be Less Identical than Identical Twins

Another reason cloned humans seem so unacceptable is the fear that they really would be carbon copies. One response, repeatedly heard this past year, is to point out that genetic clones already exist, namely identical twins. But an even better argument is that identical twins are not really identical—and that clones would be even less so. The reason is that there are more than genes at work in development, as a July report in *Nature* by genetic epidemiologist Bernie Devlin of the University of Pittsburgh Medical School and his colleagues underscored.

The team examined a long-standing puzzle: Why do fraternal twins who were raised apart from birth tend to be more alike in IQ than ordinary siblings who were also raised apart? After all, fraternal twins and sibling pairs share the same degree of genetic resemblance. Devlin concluded that one strong environmental influence had been ignored: the intrauterine environment which the twins share and the siblings do not. Just as a river cannot be stepped into twice, it appears that a mother's intrauterine environment changes with time.

But a clone and its parent will probably not even develop in the same mother, let alone in the same uterus at the same time and in the same egg. Like identical twins, they will share nuclear DNA, but otherwise they will have far less in common than twins do. Sheep are not the animals in which to investigate all this—they tend to be so alike that even their owners have great difficulty telling them apart. If humans are eventually cloned, though, it seems that we will not find ourselves in a world of ditto-heads. Instead, we will discover just how different clones can be.

12

Cloning Offers New Hope for the Childless

Virginia Morell

Last February, when Brigitte Boisselier, a French chemist, heard that Scottish scientists had produced Dolly, a sheep cloned from an adult cell, she was one of the few researchers whom the news did not suprise. A member of a fringe religious organization called the Raelian Movement, Boisselier had expected such a development: the group's leader, Rael, had predicted it 23 years before. It seems that Rael, a former French sports journalist, received the news of the impending discovery from extraterrestrials. They send him such announcements periodically, since he's half E.T. himself. According to a Raelian fact sheet (which could also serve as a script for *The X Files*), his mother was transported aboard a UFO, where she was inseminated by one of these otherworldly beings. In 1946, Rael was born "from this union," and 27 years later he began receiving messages from the distant paternal side of his family. Most of these celebrate science and technology, predicting a future when we Earthlings will "rationally understand (our) origins" and begin making synthetic people. Cloning human beings, apparently, is one of the steps we must take on this path.

"Rael told us this would happen," says Boisselier, "so when we heard the news we weren't shocked; we were organized." Indeed, so organized that one month later—even as medical ethicists, politicians, and pundits debated whether the technique should ever be applied to humans, and President Clinton asked for a moratorium on such research—the Raelians launched a company called Valiant Venture Ltd., the world's first human cloning firm. Advertised on the Web, Valiant Venture offers a service called Clonaid to help parents who want to have a child cloned from one of them. Boisselier signed on as the firm's scientific director and is now busy overseeing experiments that she believes will lead to the

---

This appeared as "A Clone Of One's Own" in *Discover*, v19, 1998.

first cloned human in a mere two years.

"We need to do many experiments first with other species to be sure that it can be done without causing any damage," says Boisselier. "And we also need to raise more funds." Nevertheless, the company, now 14 months old, is making "good progress." As of late February, it had a list of more than 100 people (Raelians and nonbelievers) who would like to be cloned or to have someone they love cloned—for a minimum fee of $200,000. Boisselier claims that her firm's research is advancing, although she would not say where the studies are taking place or who is doing them, making it impossible to verify her claims. But because the procedure can be performed in a relatively simple, inexpensive laboratory, as other scientists have noted, there is also no reason to doubt that the Raelians are doing exactly what they say: taking the first experimental steps to produce a human clone. "We've subcontracted the work to labs where it's legal to do this," Boisselier explains, noting that human cloning is banned in France. "To say that human cloning is forbidden won't stop the science," she says. "It's important that society knows that this is possible, that it can be—and will be—done ... . In a few years, I expect there'll be a lot of cloned people, that it will be done everywhere in the world. This is what happens with technological advances."

Boisselier's outspoken enthusiasm for producing human clones is rare among scientists. Since Dolly's appearance, only one other researcher—Richard Seed, a Chicago physicist turned biologist—has jumped publicly into human cloning. He held a press conference in early January to say that he intends to open up shop as soon as he raises the funds. Like Boisselier, he has a list of people who want to be cloned (although his is shorter, only four candidates), and he also thinks human cloning can be a reality in a rather short time and with only a few million dollars for start-up costs. But most other researchers are far more cautious, especially since they have yet to clone an adult of any of our closest relatives, other primates. These researchers regard announcements like Seed's and Boisselier's as not only premature but off the wall. More than one referred to Seed as a kook, an oddball simply out to make a name for himself. Seed's announcement that human cloning was part of God's "plan for humankind (to) become one with God" did not help that image.

For all their faith in science and their apparently more

rigorous approach to cloning, Boisselier and the Raelians are obviously far outside the mainstream. Their offer also plays on the fears of parents, says Mark Sauer, a reproductive endocrinologist at Columbia, since they propose to store the cells of living children. These cells could be used later to produce a clone of the child should the child die. "That's exploitation of the worst kind," says Sauer. "It plays on every parent's fears. And then what about a child who's produced that way? Will he or she be burdened by the memories of the first child?" Sauer adds that he suspects "in time, it will be possible to use adult cells to clone someone." But because of the many unanswered questions—both technical and ethical—human cloning "has not been endorsed by anyone, and certainly not by those of us working in reproductive medicine. It's premature to make these kinds of announcements and may lead to unwanted legislation." Indeed, as of late February, California had already banned human cloning, 24 other states were considering such laws, and eight bills were being weighed in Congress. Or cloning may be regulated by the Food and Drug Administration, which has asserted its right to do so.

Yet because cloning offers a way around certain reproductive problems—primarily by giving an infertile or homosexual couple a chance to have a biological child—most researchers agree that one day it's likely to be an option at many fertility clinics. Human cloning, as horrific as the idea sounds to some, will happen, they say, perhaps not as soon as Boisselier and Seed estimate but far sooner than one would have guessed before Dolly trotted onto the world's stage. "It's no longer in the realm of science fiction," says Lee Silver, a Princeton geneticist and the author of Remaking Eden, a book about cloning and other reproductive technologies. "The technological breakthrough has already happened, although the details of how to do this with human cells still need to be worked out. Once they're refined, it'll be just a matter of time."

Those refinements are already taking place. In January, scientists from a Massachusetts firm, Advanced Cell Technology, showed off three cloned calves, Charlie, George, and Albert, which were apparently produced via a more sophisticated (and patentable) technique than the one used to produce Dolly. At human fertility clinics, researchers are pursuing studies of human eggs that could lay the groundwork for cloning, although that is not the purported intent. And the National Institutes of Health has funded two projects to clone rhesus monkeys, although only embryonic and fetal cells, not those from adults, will be used. Still, these types of studies bring human cloning closer to reality.

Good old-fashioned curiosity is pushing the field as well. "Ethics aside, I have to say as a scientist I find the technological problems fascinating," says David Ledbetter, a human geneticist at the University of Chicago, voicing a sentiment others in human reproductive biology share. "Why is this difficult to do? What will it take to make it work? How do you make a clone?"

As Ledbetter's queries suggest, making a human clone is not simply a matter of following a recipe. The journal article announcing Dolly's birth didn't spell out a formula for cloning mammals; in fact, it didn't identify the actual cell that supplied Dolly's genetic material. Yet even without that key piece of information, Dolly's appearance was utterly astounding, since most biologists believed that it was impossible to produce a cloned mammal using any adult cell. "That's what everyone thought," says Don Wolf, a senior scientist at the Oregon Regional Primate Research Center in Beaverton, who's overseeing the rhesus monkey cloning project. "But Ian Wilmut (the Scottish scientist who led the Dolly project) came up with a clever innovation, a neat trick that proved us all wrong."

Before Dolly, researchers thought that adult cells could not be induced to produce a clone because they are already differentiated. As a fertilized egg develops into an adult, it divides into two, then four, then eight identical cells. Soon, however, the cells begin to specialize, becoming bone or skin, nerve or tissue. These differentiated cells all share the same DNA—the blueprint of the body—but they follow different parts of the instructions it contains. "In a sense, they're programmed," says Wolf, and as they age, it becomes more and more difficult to reprogram them, to make them switch functions. That's exactly what the Scottish team did when they produced Dolly: they took the genetic material from a differentiated adult cell and made it behave like the genetic material in a newly fertilized egg. Their success, however, does not mean that it is now easy to reprogram a human adult cell. If anything, notes Wolf, researchers suspect that every species is unique in its requirements for setting its cellular clock back to zero.

Low-key and soft-spoken, Wolf stepped into the cloning spotlight last year, when the primate center announced that he had produced two monkeys, called Neti (an acronym for "nuclear embryo transfer infant") and Ditto, using a technique similar to the one used to make Dolly. Despite Ditto's name and stories in the press, the monkeys are not identical copies of each other; they are only brother and sister. They were cloned using cells taken from two different embryos that shared an egg donor and a sperm donor. Still, their existence demonstrates that the formerly unthinkable is doable—and with primates.

Further, Wolf suspects he could produce clones from adult monkey cells as well, although, he is quick to add, he's not attempting to do so. "I have no desire to compete with the Richard Seeds of this world," says Wolf. "Nor do I want to see a knee-jerk reaction from our legislators that bans everything we're trying to do, particularly with techniques that have such tremendous potential for biomedical research." Already Congress has made research on human embryos off-limits to anyone receiving federal grants. Those who do not comply will have their labs shut down. It's safer, Wolf and others say, simply to avoid the subject.

Wolf retired from the primate center in 1996; he came back only after receiving the two NIH grants to produce a series of cloned monkeys for medical research and is now setting up his new lab. When complete, it will occupy three rooms in one of the center's squat beige buildings. In one, two researchers dressed in lab coats are peering through their microscopes at petri dishes filled with pinkish masses of monkey tissue. Somewhere in the gelatinous mix are the eggs, each about five-thousandths of an inch across. The researchers' task is to pluck out the good ones gently with a thin glass tube called a micropipette, then place these in a fresh dish for later use. Judging from the back and neck stretches the duo indulge in during a break, their efforts require almost as much concentrations as trying to induce a spoon to bend. "This is going to be the main place of activity, Room 003," says Wolf, pausing briefly to check on his group's progress, then leading the way outside, where tall pines and firs tower overhead.

Wolf moves through the lushly landscaped grounds to a nondescript conference room, where he pulls up a chair and begins explaining the enormous boon genetically identical monkeys will be to medical researchers trying, for example, to develop an AIDS vaccine. "They'd be an ideal model system," says Wolf, since they'd have identical immune systems, eliminating an important potential cause of confusion when scientists test such a vaccine or other treatment.

The center already raises rhesus monkeys for medical research; most are used in experiments here and some are sold to other medical research institutions. While awaiting their fate, the monkeys live in grassy two-acre enclosures where they pick at the grass, climb tree stumps, play, and mate, keeping an eye out for their feeders. From a distance and to the unitiated, they all look so much alike they could easily be clones. When mature, Neti and Ditto will join one of these troops. For the time being, they're kept with other young monkeys in a smaller yet roomy cage, although no one seems sure which cage they're in or if they're even in the same one. Since their brief moment of celebrity, they've been treated like any other adolescent monkeys at the center, and since they apparently look like the other adolescents, they are no longer singled out for show.

To make Neti and Ditto, Wolf followed a procedure that has frequently been used in the cattle industry for producing prized breeds. It is not, however, easily done; even in cattle, only one to four percent of such pregnancies yield offspring. In light of that low percentage, Wolf's first efforts represented "a tremendous success," says Dee Schramm, a reproductive physiologist at the Wisconsin Regional Primate Research Center in Madison. From 52 transplanted embryos, Wolf produced two healthy monkeys. "Yes, that's encouraging," Wolf acknowledges, "but you really can't draw any conclusions or expectations from what we did. After all, we've only done it twice." Still, the success has encouraged Schramm and his colleague David Watkins to attempt to clone rhesus monkeys themselves. They, too, have an NIH grant for work using fetal and embryonic cells. They also plan to produce monkey embryos from adult cells, to study the differences in embryonic development among clones produced from different sorts of cells. However, they won't implant embryos produced from adults into female monkeys; only clones made from fetal and embryonic cells will be carried to term. Schramm says he hopes to have "several pairs of identical monkeys over the next two years." The work is so slow and tedious, he adds, that "I don't foresee any monkey cloning factories."

That's because it's tricky to reprogram any differentiated cell, whether embryonic or adult. To turn a cell's clock back to zero, researchers like Wolf and Schramm use a technique called nuclear transfer technology. This is the basic method that produced Dolly, Neti and Ditto, and the three identical calves. In all three cases, the scientists removed an egg's nucleus (that is, its DNA, the genetic material that makes each individual unique) and replaced it with the nucleus from another cell. For Dolly, the nuclear material came from an adult cell; that of Neti and Ditto came from two separate embryos; and the calves' was derived from the cells of a single fetus. In all cases the cuckooed eggs were then persuaded to grow and divide normally.

That's the straightforward part of the formula. In between lies a minefield of potential problems, many unique to whatever species is being cloned. "We're not following a recipe," says Wolf. The conditions under which the embryos grow vary widely: each animal has its own required temperature, for example. And an embryo's cells begin to differentiate at different moments for different species. Sheep, calves, monkeys, and humans all reach the eight-cell stage before they start differentiating, but mice begin when the embryo consists of only two cells, which is why no one is cloning them. "There's also a lot of variation among mammalian species just in the size and nature of the egg," Wolf adds. In some mammals, such as pigs, eggs are dark in color, making it hard to tell if they are viable. While that's not a problem for manipulating rhesus monkey or human eggs, where any discoloration means the egg is dead, simply getting the eggs is. "You can get buckets of eggs from slaughterhouses" for livestock species, explains Schramm. "But every egg you get from a monkey is worth its weight in gold."

In the case of Neti and Ditto, eggs were first harvested from several rhesus females whose ovaries had been stimulated with hormones. "You give them hormone shots twice a day for eight days," says Schramm, and "then, if you're lucky, maybe you get 20 eggs. Out of these, 16 may be mature. And from these 16, perhaps 12 will be fertilized." The eggs are fertilized by placing them in a dish with the male monkey's sperm, and the resulting cells are grown in a nutrient broth under what Wolf terms "well-defined conditions; this is something we know a lot about from human infertility studies and that can be applied to our monkeys." Each embryo is allowed to grow for three days, dividing into eight cells. At this stage, all the cells are still identical to one another and largely unprogrammed. "Theoretically, you could produce a complete individual from each of these cells," says Wolf, giving you eight identical monkeys. But only theoretically, because most do not survive the coming manipulations.

In the next step, the cells, called blastomeres, are carefully teased apart; they constitute the donor nuclei. "Each one," explains Wolf, "is really one-eighth of an embryo," but that one-eighth contains the key ingredient: the nuclear DNA, all that's apparently needed to get the process ticking again.

You might expect that geneticists could divide each embryo into eight blastomeres, wait for each blastomere to grow into an eight-cell embryo, and repeat the process indefinitely. But that's not possible, says Wolf, because the embryo's cells begin differentiating into limbs and organs after a certain amount of time has passed since its development began, regardless of how many cells it has. An embryo grown from a blastomere will have only an eighth as many cells to work with as an entire embryo; if you divided it again, it would have only a sixty-fourth as many cells. "As development proceeds, when time for it to differentiate arrives, it doesn't have enough cells for the job," says Wolf, and even a blastomere will be less viable than an entire embryo. Because the cues to develop come from the cell's cytoplasm—the material that fills the cell—rather than the nucleus, the blastomere's clock can be reset by transferring its genetic material to a new egg full of fresh cytoplasm.

Using micropipettes, the scientists remove and discard the nuclear DNA from another batch of rhesus monkey eggs. That leaves the cytoplast—that is, the egg's membrane and the material that once surrounded its chromosomes. A donor cell, one of the blastomeres, is then placed next to the chromosome-free egg in a petri dish. "In normal fertilization, an egg is in a quiescent state at the time it is ovulated," says Wolf. "The sperm triggers the egg to be activated, and the cytoplasm starts the program of events that will lead to development. But here, we aren't giving the cytoplast any sperm, so we must artificially stimulate" the two cells with a chemical treatment. A pulse of electricity then causes the two cells to fuse, and a "reconstituted" embryo is formed. The order of these two steps, however, was reversed when the Massachusetts researchers at Advanced Cell Technology cloned their calves; and the chemical treatment was apparently bypassed altogether when Dolly was made. "It could be species differences, or it could be artifacts of the lab," says Wolf. "It's too early to say.

"Once we have the embryo, we can treat it as we do any other," he continues. "Most often we freeze them until we have a monkey ready for an implant." That's the other big hurdle—making sure that the recipient monkey is at the right point in her cycle for the embryo to take. Prospective recipients are monitored for several weeks beforehand. To do the actual transfer, a veterinarian surgically places the embryo into the monkey's oviduct. "Women have short, straight cervixes," explains Wolf, "so surgery isn't required" when embryos are transplanted at fertility clinics. "But a monkey's cervix is tortuous, and the only way we can implant the embryos is surgically, although we're trying to come up with other methods."

At the end of all this labor, only eight twins can be produced, and that's assuming that every transfer succeeds, which is "pie in the sky," says Wolf. "It's not the optimal method, although we used it to make Neti and Ditto." But Wolf wants a series of clones, and for this, he says, "we need a lot of identical nuclei." He expects to retrieve these donor nuclei from the cells of fetal monkeys, such as their embryonic stem cells (undifferentiated precursors for other cell types) or fibroblasts (the cells that form the body's connective tissue, which are commonly grown in labs). Both kinds can be propagated in large numbers in test tubes, making it possible, he says, "to produce a clone size that is infinite in number." In other words, he expects to turn out identical monkeys, like a copying machine with a jammed "on" switch. "We don't know yet if we can do this; that's what we're working on now."

And in fact, this same technique—growing a line of fetal cells for subsequent nuclear transfer—enabled researchers at Advanced Cell Technology to produce the identical calves. "It's a very efficient method for us already," says Steven Stice, the firm's chief scientific officer. "We're producing more viable embryos than we have cows to put them in." (Oddly, the company has had no luck cloning a pig. "They are very different from cattle," says Stice. "Every step has to be reevaluated. We're not sure what we're doing wrong.").

Scientists first began trying to clone animals using adult cells in 1938, when the German embryologist Hans Spemann proposed making a clone by removing an egg's nucleus and replacing it with the nucleus from another cell. Those efforts failed until the 1970s, when frogs were finally cloned via the nuclear transfer method. None of the cloned frogs, however, made it past the tadpole stage. And that's where the idea of adult cloning stayed until Dolly arrived.

"It couldn't be done; that was what everyone said," explains Wolf, "which is why this was such a revolutionary discovery." The Scottish team "found a way to reprogram that adult cell." They did so by starving the adult cells, thus inactivating them. Wilmut began with a vial of frozen cells taken from the udder of a six-year-old sheep. His team thawed them and placed them in a growth serum with only minimal nutrients for five days. "That's the trick that made all the difference," says Wolf. The adult cells were then fused with 277 different eggs. Out of all these attempts, only one lamb was born: Dolly. "That tells you that something was desperately wrong with the other 276," says Steen Willadsen, an embryologist at St. Barnabas Medical Center in Livingston, New Jersey.

Because of this low success rate, "we're a long ways off from getting adult cloning to work on a regular basis even in domestic animals," adds Lawrence Layman, a reproductive endocrinologist at the University of Chicago. "It'd be highly unethical at this stage to try it in humans," since the probability of miscarriages and birth defects is high. Stice agrees: "It'd be complete folly. We've used hundreds of thousands of eggs in cattle over the last ten years to achieve these results. To start at ground zero now with humans would be morally wrong and misguided."

Some researchers worry too that damage from aging DNA may be passed on to the cloned infant. "It's going to be very instructive, watching Dolly age," says Julian Leakey, a biochemist at the National Center for Toxicological Research in Jefferson, Arkansas. "If she goes through puberty, she may be okay." But she might also have acquired some random genetic mutation that could lead to problems early in life. "That's the potential danger of cloning adults," says Leakey, "which is why it would be useful to do controlled tests in short-lived mammals, such as rodents, first. Then you could work out the odds of using aged tissue versus young tissue for cloning."

There are other unanswered questions. It's not clear which cells were used to make Dolly. "They don't know which cell from the udder worked, or why it worked," says Ledbetter. "That's a big gap, and it means we don't have any idea if every cell type will work or only certain ones." Some researchers even question that it was an adult cell at all: the udder cells were taken from a pregnant sheep, and fetal cells are known to circulate in a mother's body.

Nor do researchers know if the serum starvation trick will work with other species.

Despite the difficulties, says Willadsen, "the technique will be—is being—perfected" ... somewhere. And once that happens, it's only a matter of time before we see the first cloned humans—individuals who are a physical copy, or twin, of their mother or father, but separated by at least a generation. "When that first cloned child is born, not only will no one know that he or she is different," says Lee Silver, "no one will know that he or she is a clone. People will probably say things like, 'Oh, you look so much like your mother (if she was the nucleus donor), and she'll smile. But no one will know, at least not until the kid is 16 and decides to sell her story to the tabloids for a million dollars."

From studying twins that were separated at birth and raised in different families, researchers surmise that such clones will also be likely to share intellectual abilities and personality traits with their sole biological parent. Clones may thus follow in the footsteps of their parent—but only in a very general way. "They will be separated by an entire generation," notes Sandra Scarr, a professor emerita of psychology at the University of Virginia in Charlottesville. "And as we all know, the cultural and social circumstances of the next generation are never the same as those of the preceding one. It's those social attitudes that shape a person's view of the world, including everything from how you view the stock market to the excesses of war. So the clones may be similar in intellect and personality, but their content will be different." Identical twins reared in the same house, she notes, listen to the same bedtime stories, eat at the family table together, attend the same schools, have the same friends and teachers. The clone and its parent, however, will share none of these experiences. "And these are the kinds of things that influence how one expresses one's genetic potential."

"People think it's going to be a robot or automaton," says Thomas Bounchard, who's led the long-term twin studies at the University of Minnesota. "Nothing could be further from the truth. They'll be their own persons, and that's why the idea of cloning doesn't bother me in the least. It's nonsense to be afraid of it." Yet because of this culturally ingrained idea of what a clone is, some ethicists are concerned that the parent of a clone may try to exert excessive control over the child. "Parents already control their children to an extraordinary degree," says Lori Andrews, a law professor at the Chicago-Kent College of Law. "Will these clones be held in some kind of genetic bondage to their parent? They might put undue pressure on the child to grow up in a certain way, so that it really doesn't have its own identity."

Other researchers question how similar the clones will be, even physically. "We already know from studying monkeys and children that there's considerable variation at birth," says Christopher Coe, a psychologist at the University of Wisconsin in Madison. Coe intends to explore this variation with the cloned rhesus monkeys that his colleagues Watkins and Schramm are attempting to produce. Since the cloned embryos will be implanted in different mothers, they'll "give us the best opportunity we've ever had to clarify what we mean by nature versus nurture," says Coe. "It's the project I've dreamed about since graduate school, 20 years ago." For instance, how different will a cloned monkey that's implanted in an older mother be from one that's grown in a younger mother? "To what degree do in utero influences affect the development of the baby?" asks Coe. "And how much do the mother's actions, what she eats, and whether or not she's dominant or submissive, influence her baby's growth? The prenatal environment plays a far bigger role in shaping a baby than most people realize." The cloned monkeys, he believes, will probably look alike (although they could also differ in such things as their weight at birth) but will nevertheless "be quite different."

Other research on human twins also suggests that such things as how early the cells divide into twins and where the twins are placed in the uterus affect how "identical" they are after birth. "I think that's the real question: Just how different will these cloned babies be at birth, despite being genetically identical to their parent?" adds Coe. In an effort to establish the cloned monkeys' individuality, he will be measuring everything from their birth weight to how quickly they hold up their heads and how long they nurse.

Then too there's the question of the influence of the mitochondrial DNA. Not all of a cell's DNA is found in the nucleus; the mitochondria, tiny organs a cell uses to transform food into energy, have their own DNA. Although the donor egg will receive a new nucleus, it will retain its mitochondrial DNA, which may well be different from the donor's. "It's only a small amount of genetic variation, but it's there," says Silver, though, he adds, "there is nothing in the mitochondrial DNA that matters in making us different from each other."

In short, cloning yourself will not roll the clock back. It will not produce your soul mate and may not even give you your complete identical twin. What it will do is give you a baby that is more biologically related to you than anyone else. And that, says Silver, is why cloning will happen and few people will harshly judge those with infertility problems who choose it as a way to reproduce. "It's instinctive, I think, to want to have a biological child. That's what cloning offers—a chance for some people to have what they thought they never could have: a child of their own."

III

The Issues in Brief

Introduction to Part III

In Part III we begin to look directly at some of the contentious areas relating to cloning. Charles White (Reading 13) sharply counterposes the clear benefits human cloning may offer some people with the difficult ethical problems raised, and suggests that public polarization over cloning may soon come to exceed the current passionate culture-war between pro-choice and pro-life advocates.

In Reading 14, Jim Motavalli and Tracey Rembert report on views expressed, chiefly against human cloning, by a number of spokespersons. *National Review* (Reading 15) gives us a clear statement of the view that some of the motives for cloning are "deformed morally." Charles Krauthammer calls for the immediate and permanent banning of human cloning (Reading 16). Readers can decide for themselves whether this is consistent with his view of a few months earlier (elsewhere in this volume) that no prohibition of the cloning of humans could be effective. Michael Gough of the Cato Institute holds (Reading 17) that hasty anti-cloning legislation might hamper research that could provide benefits. Reading 18, by Eliot Marshall, identifies some of the interest groups who mobilized to oppose prohibition of cloning by the United States federal government

13

The Controversy over Cloning May Become More Furious than the Abortion Debate

Charles A. White

In the past 40 years, scientists have cloned (made an exact replica of) frogs, mice, cattle, and sheep. It was inevitable that researchers would prove the possibility of cloning humans, and last October geneticists Jerry Hall and Robert Stillman of George Washington University did exactly that.

Life starts with the embryo, the mother's egg fertilized by the father's sperm. During the Hall/Stillman experiment, development began when the embryo divided into two identical cells. While quickly isolating these two cells, the researchers were forced to strip away their outer coatings but replaced them with an artificial gel. Protected in this way, the microscopic cells began to reproduce. Using this technique many times, Hall and Stillman produced 48 clones in all.

The purpose of the research was to help parents who have turned to *in vitro* fertilization (IVF) because for some reason they can't achieve a normal pregnancy. If the woman has only one embryo, the chances of a successful pregnancy through IVF are small, but adding three or four cloned embryos would greatly improve the odds. This was the aim of the Hall/Stillman research, and none of the embryos they cloned were allowed to grow for more than six days.

News of their experiment, however, flashed around the world with the speed of light, and critics wasted no time in expressing their horror. People predicted the sinister cloning of another Adolf Hitler or Jurassic-Park–type cloning in which DNA (the basic material in the gene) is preserved and grown into a

This appeared as "Ethical Debate" in *Canada and the World*, v59 (Jan. 1994). pp. 8–9.

living replica of the original. The mildest criticism was that cloning would destroy human individuality and dignity. A *Time* Magazine/CNN poll indicated that three out of four Americans disapprove of human cloning and that 40 percent would put a temporary stop to all research.

However, fantasies about recreating dinosaurs, another Hitler, or even another Einstein belong to the distant future. The human embryos in the recent experiment were immature and had not yet begun to develop into the specialized cells required for skin, bones, and the various organs of the human body. The information which will lead to this specialization in the adult cell is present in the single cell, but scientists so far have no access to this information or how it results in specialization.

Even so, human cloning does raise ethical questions which are certain to be fiercely debated in the years ahead; medical ethicists and other moralists are already going at it hammer and tongs. Apart from the original goal of the researchers to help infertile couples achieve an *in vitro* pregnancy, here are examples from a wide range of possible future uses of cloning:

A baby girl is born by IVF, grows up, is implanted with a clone of her own embryo and becomes the mother of her twin sister.

A woman knows she will soon become sterile because of chemotherapy or exposure to a toxic substance. She clones an embryo now for possible future use.

Spare embryos could be sold to potential buyers at values which might fluctuate as the buyers watched to see how well the twin already born was turning out.

A couple know that their children run the risk of developing hemophilia or sickle-cell anaemia. Medical scientists can now screen for such diseases but the process sometimes kills the embryo. Having an extra clone or two on hand might make the difference between passing on a defective gene or producing a perfectly normal baby.

An IVF clinic could offer prospective parents a catalogue of children's photographs under each of which would be an outline of the child's academic and social performance. A couple could choose from the list, receive a frozen embryo, and raise a child who would be an exact replica of the child pictured.

A couple could hold in reserve a cloned embryo of each of their children. If a child died it could be replaced with a genetic duplicate. If another needed a bone marrow or kidney transplant, a cloned embryo could be raised to provide a perfectly compatible transplant.

Critics call such scenarios "playing God" and interfering with an individual's right to his or her unique genetic blueprint. On the other hand, cloning offers possibilities for fighting disease and relieving human suffering. Arthur Schafer, director of the Centre for Professional and Applied Ethics at the University of Manitoba, suggests that "embryo cloning ... to secure fetal brain tissue ... shows great promise in the treatment of Parkinson's and other hereditary diseases." Abortions now provide an adequate supply of tissue, but with promising new research on Alzheimer's disease "demand may soon skyrocket ... and cloning may permit the tissue to be produced *in vitro* ... "

There are now 10,000 frozen embryos in the U.S. waiting in refrigerated limbo for would-be parents to come along, so the human tendency to play around with Mother Nature is already far advanced. Human cloning makes answers to questions of right or wrong, benefit or harm, even more urgent. Its possibilities stretch all the way from dubious tinkering with family relationships to curing hereditary disease. Government regulations or at least guidelines for cloning would seem a reasonable way to exert some form of control.

Canada has no official position on human cloning and no legislation is in sight. The United States has only a jumble of state laws and local review boards (one of them at George Washington University approved the Hall/Stillman research). However, it's likely that some kind of national watchdog will be appointed during President Bill Clinton's term.

Looking further afield, we find that the recent U.S. experiment would punish the researchers by up to five years in prison in Germany. British law requires a license to clone human cells which the authority concerned refuses to grant; the penalty for violation is up to 10 years' imprisonment. Guidelines have the force of law in Japan and ban all research on human cloning. In all, more than 25 countries have commissions that set policy on all kinds of reproductive technology.

In coming years, the angry standoffs on abortion between Pro-Life and Pro-Choice groups may seem mild compared to the polarization of society over human cloning.

14

The Balance of Opinion Is Against Cloning

Jim Motavalli and Tracey C. Rembert

In the recent film *Multiplicity*, the condo contractor played by Michael Keaton discovers that having a clone of himself around to help with the heavy lifting is not quite the trouble-free experience he had imagined. So too, in real life, is the prospect of human cloning rather daunting.

While human cloning has been the stuff of science fiction for centuries, the possibility that it actually could happen is of much more recent vintage. In July of 1996, Dr. Ian Wilmut of the Roslin Institute in Edinburgh, Scotland became the first scientist to successfully clone and adult mammal. (The cloning of embryos has been going on much longer, with Dr. John Gurdon cloning an adult South African tree frog in 1962.).

It took Dr. Wilmut 277 tries to create the lamb Dolly from the merger of a six-year-old sheep's mammary cell with a prepared sheep's egg. Since Dolly met the press, a number of scientists have come forward to acknowledge that they, too, are experimenting with the cloning of mammals. Their interest is not purely scientific. Wilmut's Roslin Industries, for example, is in partnership with the biotechnology company PPL Therapeutics, whose aim is to sell cancer- and cystic fibrosis-fighting human proteins collected from genetically engineered animal "hosts."

A U.S. company, Alexion Pharmaceuticals of New Haven, Connecticut, is working to develop a "super" pig whose organs (genetically coated with human proteins to minimize cross-species rejection) would interchange with those of humans. But this raises the specter of animal viruses like "mad cow" disease (bovine

This appeared as "Me And My Shadow" in *E: the Environmental Magazine*. v8 (July/Aug. 1997). pp. 15–16. Contact: Union of Concerned Scientists, 26 Church Street, Cambridge, MA 02238. 617-547-5552.

spongiform encephalopathy) crossing over to humans along with the transplants. "One of the things people have learned is that it's not wise to cross such boundaries," says Dr. Margaret Mellon, a scientist with the Union of Concerned Scientists (UCS) who serves on its panel investigating the ecological risks of genetically engineered products. Despite the warnings, there are more than 190 genetically engineered animals awaiting patents by researchers and corporations.

Is it a short step or a giant leap from a cloned sheep to a cloned human being? Few scientists doubt that human cloning is possible. "It seems very likely that it would work with humans," says Brigid Hogan, a biologist at Vanderbilt University. *Nature* magazine, which first published Dr. Wilmut's work, estimates that adult human cloning is likely to be achievable "in one to 10 years."

A human clone will never be a perfect copy, says Dr. Glenn McGee, an assistant professor in the Center for Bioethics at the University of Pennsylvania and the author of The Perfect Baby. Parents who would consider cloning a dead child, for example, should realize "that it would not be the same child, though it would look very similar. Environmental factors shape who we are."

The prospect of human cloning has obvious ethical dimensions, and it was for that reason that President Clinton (who banned the use of government funds for human cloning research last March) asked a federal panel to look into the matter and make recommendations. Dr. Alta Charo, a professor of law and medical ethics at the University of Wisconsin, sits on that body, the National Bioethics Advisory Commission (NBAC). Dr. Charo notes that only ten states have laws that regulate human bioengineering, and that all of them only "address the forming of embryos from the union of egg and sperm."

Taking note of the vast media interest in the subject, Dr. Charo cautions, "The research is still so enormously preliminary. We've yet to see Dr. Wilmut's success replicated in a single sheep, let alone in people."

Also on the federal panel is Dr. Thomas Murray, director of the Center for Biomedical Ethics at the Case Western Reserve School of Medicine in Cleveland. He says that the panel has talked to a group of theologians representing the Protestant, Catholic, Jewish and Islamic faiths. "None of them said it was a wonderful thing," Dr. Murray says, "and many are concerned that it's an exercise in human hubris."

Hubris is putting it mildly, says bioengineering critic Jeremy Rifkin. "This is the most radical experiment conducted against human nature in history," he says. "We have here a new eugenics movement that goes beyond Hitler."

Even without the human dimension, cloning presents ethical dilemmas. Dr. Jane Rissler, a Washington-based UCS senior staff scientist in agriculture and biotechology, is concerned that cloning will further weaken the already imperiled genetic diversity of agricultural crops and animal species. "Both crops and livestock have become very uniform and much more vulnerable to disease," she says.

But while cloning poses a threat to some plants and animals, it could also be used to save others, says Dr. Betsy Dresser of The Audubon Center for the Research of Endangered Species in New Orleans. Dr. Dresser, who headed the team that created the first test tube gorilla at the Cincinnati Zoo in 1995, says that cloning could help preserve endangered animals. "Theoretically, what worked with Dolly could work with any mammal," she says. For example, a wild-caught orangutan could be cloned and his "copies" used to enrich the genetic diversity in a number of zoo populations.

While cloning animals has received tentative public approval (according to ABC News, 53 percent of Americans approve of cloning animals for medical research) human xeroxing is decidedly less popular. A full 87 percent oppose cloning humans, and 93 percent don't want to be cloned themselves.

15

Cloning Humans Is a Slippery Slope

By the editors of *National Review*

Eventually someone, if not Dr. Richard Seed, will be able to open a clinic for cloning human being. When that happens, let it be in a back alley. Human cloning should be banned not for five years or ten years, as President Clinton and Senator Dianne Feinstein (Democrat–California) respectively urge, but always. Deliberately to create human beings for experimental purposes, or for spare parts—or to satisfy someone's vanity—is as deformed morally as the initial results are certain to be physically. Libertarian writer Ronald Bailey argues that having the same genes as another person would not give human clones an inferior moral status, any more than it does identical twins: "Kids are not commercial property—slavery was abolished some time ago." But why should they not be treated as property when they are manufactured rather than begotten? Whence will flow this product's claim to dignity and rights?

It is true that we have already begun to sever reproduction from sexual intercourse, and Laurence Tribe warns us that to ban cloning would endanger such noble causes as surrogate motherhood and gay adoption. Cloning would, however, be a fateful step from redefining parenthood to redefining the human race. The bans proposed by Reps. Dick Armey (Republican–Texas) and Vern Ehlers (Republican–Michigan) may be overly broad; such bills could forbid the cloning of discrete organs, which does not have the same moral drawbacks, and which could have serious medical benefits. But cloning human beings is a slippery slope, the top rung of which is already in the depths.

Published as "Seeds of Trouble" by the editors of *National Review* in *National Review*, v50 n2, pp.19, Feb 9, 1998.

16

Creation of Headless Humans Should be Banned

Charles Krauthammer

The ultimate cloning horror: human organ farms.

Last year Dolly the cloned sheep was received with wonder, titters and some vague apprehension. Last week the announcement by a Chicago physicist that he is assembling a team to produce the first human clone occasioned yet another wave of *Brave New World* anxiety. But the scariest news of all—and largely overlooked—comes from two obscure labs, at the University of Texas and at the University of Bath. During the past four years, one group created headless mice; the other, headless tadpoles.

For sheer Frankenstein wattage, the purposeful creation of these animal monsters has no equal. Take the mice. Researchers found the gene that tells the embryo to produce the head. They deleted it. They did this in a thousand mice embryos, four of which were born. I use the term loosely. Having no way to breathe, the mice died instantly.

Why then create them? The Texas researchers want to learn how genes determine embryo development. But you don't have to be a genius to see the true utility of manufacturing headless creatures: for their organs—fully formed, perfectly useful, ripe for plundering.

Why should you be panicked? Because humans are next. "It would almost certainly be possible to produce human bodies without a forebrain," Princeton biologist Lee Silver told the London *Sunday Times.* "These human bodies without any semblance of consciousness would not be considered persons, and thus it would be perfectly legal to keep them 'alive' as a future source of organs."

"Alive." Never have a pair of quotation marks loomed so

This appeared as "Of Headless Mice ... And Men" in *Time.* v151 (Jan. 19 1998), 1998. pp. 76.

ominously. Take the mouse–frog technology, apply it to humans, combine it with cloning, and you are become a god: with a single cell taken from, say, your finger, you produce a headless replica of yourself, a mutant twin, arguably lifeless, that becomes your own personal, precisely tissue-matched organ farm.

There are, of course, technical hurdles along the way. Suppressing the equivalent "head" gene in man. Incubating tiny infant organs to grow into larger ones that adults could use. And creating artificial wombs (as per Aldous Huxley), given that it might be difficult to recruit sane women to carry headless fetuses to their birth/death.

It won't be long, however, before these technical barriers are breached. The ethical barriers are already cracking. Lewis Wolpert, professor of biology at University College, London, finds producing headless humans "personally distasteful" but, given the shortage of organs, does not think distaste is sufficient reason not to go ahead with something that would save lives. And Professor Silver not only sees "nothing wrong, philosophically or rationally," with producing headless humans for organ harvesting; he wants to convince a skeptical public that it is perfectly O.K.

When prominent scientists are prepared to acquiesce in—or indeed encourage—the deliberate creation of deformed and dying quasi-human life, you know we are facing a bioethical abyss. Human beings are ends, not means. There is no grosser corruption of biotechnology than creating a human mutant and disemboweling it at our pleasure for spare parts.

The prospect of headless human clones should put the whole debate about "normal" cloning in a new light. Normal cloning is less a treatment for infertility than a treatment for vanity. It is a way to produce an exact genetic replica of yourself that will walk the earth years after you're gone.

But there is a problem with a clone. It is not really you. It is but a twin, a perfect John Doe Jr., but still a junior. With its own independent consciousness, it is, alas, just a facsimile of you.

The headless clone solves the facsimile problem. It is a gateway to the ultimate vanity: immortality. If you create a real clone, you cannot transfer your consciousness into it to truly live on. But if you create a headless clone of just your body, you have created a ready source of replacement parts to keep you—your consciousness—going indefinitely.

Which is why one form of cloning will inevitably lead to the other. Cloning is the technology of narcissism, and nothing satisfies narcissism like immortality. Headlessness will be cloning's crowning achievement.

The time to put a stop to this is now. Dolly moved President Clinton to create a commission that recommended a temporary ban on human cloning. But with physicist Richard Seed threatening to clone humans, and with headless animals already here, we are past the time for toothless commissions and meaningless bans.

Clinton banned federal funding of human-cloning research, of which there is none anyway. He then proposed a five-year ban on cloning. This is not enough. Congress should ban human cloning now. Totally. And regarding one particular form, it should be draconian: the deliberate creation of headless humans must be made a crime, indeed a capital crime. If we flinch in the face of this high-tech barbarity, we'll deserve to live in the hell it heralds.

17

Similar Negative Reactions Greeted Previous Advances

Michael Gough

Few biological results have excited as much attention as the announcement of Dolly's birth eleven months ago. Dolly was important and surprising because, it was claimed, she was produced from the DNA of an adult sheep.

Mammalian life begins with a "totipotent" fertilized egg that can multiply and differentiate into all the diverse types of cells—skin, nerves, bones, muscle, etc.—that make up a mature animal. As cells differentiate into specialized cells, they lose the capacity to carry out the functions of other cell types; they are no longer totipotent. A skin cell cannot produce a nerve, bone, or muscle cell, for example.

Dolly was a surprise because she was, apparently, the product of DNA from a differentiated, specialized cell from the udder of a mature sheep. The DNA was introduced into a DNA-less egg, and the egg was implanted into the uterus of a sheep where it developed into Dolly.

Dolly, at the time the experiment was announced last year, appeared to open up the possibility of human cloning. In theory, DNA could be taken from a woman or man and inserted into a DNA-less egg, and the egg, which now contained the genetic information from the donor, could be introduced into the uterus of a woman. If a child resulted from the process, she or he would be genetically identical to the woman or man from whom the DNA came.

Human cloning, if it is ever accomplished, will offer the promise of a child to love and cherish to couples who otherwise would be childless. Although cloning has been greeted very negatively, it is also true that negative reactions met almost every

---

This appeared in the *Congressional Record* on February 10, 1998, pp. S573–4.

advance in human reproduction technologies—artificial insemination, *in vitro* fertilization, "fertility drugs," prenatal diagnoses. Those technologies became accepted when they gave healthy children to couples that otherwise would have been childless.

Many scientists, including the Director of the National Institutes of Health, worry that hastily drafted and loosely drawn legislation directed against cloning will foreclose research that promises new drugs and the capacity to replace or repair nerves, skin, and muscle lost to injury or disease.

IV

Do the Benefits of Cloning Outweigh the Risks?

Introduction to Part IV

People who favor continued research on cloning tend to emphasize the tangible benefits it may bring, while opponents stress the ethical, religious, and esthetic objections, the wider social consequences, and the risks of outrageous abuse. In Part IV we turn to the benefits that may flow from further advances in cloning technology, which to some people are well worth the (arguably remote) risks.

Mark Nichols (Reading 18) focuses on the promising future applications of cloning in raising livestock. Elizabeth Pennisi (Reading 19) offers a more detailed look at animal cloning. In Reading 20, J. Madeleine Nash draws attention to some less-frequently discussed benefits and risks of cloning applied to humans, while in Reading 21 Virginia Postrel, of *Reason* magazine, underscores the potential of cloning technology to alleviate human suffering, while rejecting any appeal to pristine "nature."

Robert Winston contends that some of the risks of permitting cloning have been overstated (Reading 22). In Reading 23, Mari Jensen reports upon one intriguing line of research in which cloning technology has been used.

18

A New Cloning Technique Opens Up Future Possibilities for Livestock and People

Mark Nichols

For more than a decade, scientists have been using genetic technology to produce biologically identical copies, or clones, of animals. In theory, cloning can be used to improve sheep and cattle breeds by ensuring that the animals' most desirable genetic characteristics are passed on. But in practice, cloning has often proved disappointing: scientists have been limited in the number of clones they could produce, and the young animals frequently have a low survival rate. Now, scientists at the Roslin Institute near Edinburgh have demonstrated a dramatically different kind of cloning technology. Starting with cells from a sheep embryo, they grew thousands of copies in a culture. Technicians then fused the cells to unfertilized eggs and implanted the eggs in female sheep. In the end, only a handful of cloned Welsh Mountain lambs were born. But members of the Roslin team said that when the new technique is perfected, it should be possible to create thousands of identical sheep and cattle at a time. "This is very exciting," said Prof. Allan King, an embryologist and geneticist at Ontario's University of Guelph. "It has big implications for livestock breeding and production."

The experiment, described in the March 7 issue of the British scientific journal *Nature*, suggested that the new technology could be used someday to create cattle with leaner meat and cows that produce low-fat milk. Keith Campbell, the cell biologist in charge of the experiment, said that kind of genetic fine-tuning could become possible because, unlike existing methods, the new cloning system would enable scientists "to make

This appeared as "Send in the Clones", *Maclean's*, v109 (March 18 1996), p. 55.

much more precise genetic changes in the cells used to produce cloned animals."

The Roslin scientists scored an unexpected triumph when they achieved a type of cloning that had defeated past attempts by American and European scientists. The method differs from existing cloning technology in several important ways. In conventional sheep cloning, technicians usually remove embryos, consisting of between 50 and 60 cells, from artificially inseminated ewes, divide the cells into two clusters and re-insert these into recipient ewes.

The Roslin team started with slightly more mature embryonic cells, which were then grown in a culture where they multiplied rapidly—providing a far higher number of potential clones than usual. According to Campbell, the cells' high rate of growth may have been induced when scientists withdrew some of the nutrients in the culture. "This put the cells into a quiescent state," said Campbell, "which may have made them more suitable for controlling development into a fetus."

The use of a culture for growing cells should also make finer genetic tuning possible. In the past, scientists have tried to inject new genes into an embryo before cloning—an approach, said Roslin team member Ian Wilmut, that "is very primitive, like firing a shotgun." Using a culture, added Campbell, "we should be able to make much more precise genetic changes, and then use only the altered cells to produce new animals."

Inevitably, advances in animal cloning raise the prospect of scientists applying the same techniques to humans. Doctors at George Washington Medical Center in Washington, did just that in 1993 when they produced 48 short-lived clones of human embryos. The controversial experiment, which was made public at a scientific conference held in Montreal, triggered a fierce controversy. Since then, some industrialized nations, including Canada, have issued guidelines against the cloning of human embryos. Meanwhile, the survival rate among cloned animals remains low. Of the five Roslin lambs born in Scotland last July, only two survived infancy and were still living last week as their story was told to the world.

19

Dolly Has Stimulated Advances in Pharming

Elizabeth Pennisi

Births are usually announced on a newspaper's society or personal pages, not on the front page. But that convention didn't apply to Dolly and Polly and—just last week—George and Charlie. These white-faced lambs and Holstein calves made headlines as the products of cloning technologies that have generated fascination and fear—a reaction fanned this month by the improbable claims of a physicist who says he plans to clone adult humans within two years (*Science*, 16 January, p. 315). But the technologies have done more than spawn an ethical debate about the prospects for human cloning: They have also galvanized efforts to create transgenic livestock that will act as living factories, producing pharmaceutical products in their milk for treating human diseases and, perhaps, organs for transplantation.

That was always the main intention of Dolly's creators, Ian Wilmut, Keith Campbell, and their colleagues at the Roslin Institute and PPL Therapeutics in Roslin, Scotland. But in the year since the announcement of Dolly's birth, a dozen other groups have been adapting the technique used by the team in Scotland. Some want to clone animals bearing working copies of transplanted genes. Although key problems remain to be solved, these efforts—many of which were reported here at last week's annual meeting of the International Embryo Transfer Society—have already resulted in the birth of sheep containing a human clotting factor gene and calves containing foreign marker genes. Experiments in which the nuclei of pig cells have been fused with cow eggs have also given tantalizing results.

This work is invigorating the "pharming" industry:

This appeared as "After Dolly, a Pharming Frenzy", in *Science*, v279 n5351, pp.646–48, January 30, 1998.

Underwriting the cloning frenzy are biotech and pharmaceutical companies eager to cash in on its potential for creating transgenic livestock. "There is a huge industry that is organizing itself around [the new cloning] technology," says James Robl, a developmental biologist of the University of Massachusetts, Amherst.

There is, however, a crucial difference between these experiments and the original Dolly breakthrough—a distinction that has sometimes been lost in the public discussion of the implications of these new results. Dolly was cloned by taking nuclei from adult mammary gland cells, starving them of nutrients to reset their cell cycles, then fusing them with sheep eggs whose own nuclei had been removed. But this procedure was very inefficient—producing only one success out of the 277 eggs that took up the new DNA. The later experiments all use nuclei from fetal cells, which have proved more efficient at generating viable offspring than adult cells. Indeed, so far the Dolly experiment has not been exactly replicated, and some scientists have even questioned whether Dolly is in fact the clone of an adult.

Animal geneticists have jumped on the technology because it potentially offers a far more efficient way to produce transgenic animals than previous techniques, which involve the injection of foreign DNA into newly fertilized eggs. The success of an egg injection is not known until after the offspring is born. For example, using egg injection, PPL Therapeutics took years to develop a flock of 600 transgenic sheep, as only about 4 percent of the lambs carried the desired gene.

In contrast, nuclear transfer technology allows researchers to select as nucleus donors only those cells that express the transplanted gene. Moreover, in theory, those cells could provide as many clones as needed in a single generation. "In one fell swoop, you get what you want," says PPL research director Alan Colman. Indeed, Will Eyestone of PPL's Blacksburg, Virginia, facility told last week's meeting that egg injection "may well become old-fashioned."

Commercial Research Teams Have Made Breakthroughs

Campbell, who recently moved from the Roslin Institute to PPL's labs 300 meters down the road, Wilmut, and their colleagues were the first to announce that they had been able to produce transgenic animals with cloning technology. They reported in December that they had produced three cloned sheep, two of which are still alive, carrying the human factor IX clotting protein (*Science*, 19 December 1997, p. 2130).

Now, Advanced Cell Technology has achieved in cows what the team in Scotland did with sheep: Robl and his colleague Steven Stice announced at last week's meeting the birth of two calves carrying a foreign gene. To produce these transgenic animals, the researchers first grew bovine fetal fibroblast cells in the laboratory and then added an antibiotic-resistance "marker" gene. Only the cells that took up the gene survived exposure to an antibiotic added to the culture dishes. The researchers then fused nuclei from the survivors with enucleated cow eggs, employing a variation on the technique used by Wilmut's group. About 40 percent of the resulting embryos continued to develop once inside foster mothers, and two calves—George and Charlie—were born in mid-January. A third has been born since the announcement, and more are on the way. "[They are] the first transgenic cloned calves, and that's great," says Campbell of PPL, which is also doing nuclear transfer work in cattle. The three calves show "the phenomenon and the technology are not restricted to one species," adds nuclear transfer pioneer Kenneth Bondioli of Alexion Inc., a biotech company in New Haven, Connecticut.

That demonstration has been eagerly awaited. Transgenic cows, which produce 9,000 liters of milk per year, should be better factories for therapeutic proteins than sheep or goats. "Milk is cheap, and we have an incredible dairy infrastructure," points out Carol Ziomek, an embryologist with Genzyme Transgenics in Framingham, Massachusetts.

Indeed, that potential has already spurred a gold rush. In October 1997, Genzyme Transgenics awarded Advanced Cell Technology a five-year, $10 million contract to develop transgenic cows that will produce albumin, a human blood protein used in fluids for treating people who have suffered large blood losses. And earlier this month, Pharming Holding N.V. in Leiden, the Netherlands, formed an alliance with ABS Global, an animal breeding company in DeForest, Wisconsin, and its spin-off company, Infigen Inc., to develop transgenic cattle that produce the human blood proteins fibrinogen, factor IX, and factor VIII in their milk.

Other efforts are aimed at expanding the utility of pigs, particularly in biomedicine. A few companies and research groups hope to use pig organs or tissue to help meet the large unfilled demand for transplant organs. The goal is to genetically modify the animals' tissues so they are less readily rejected. Also, because a pig's physiology is more like a human's than is a mouse's, some animal scientists argue that pigs could be good models for studying human diseases if their genetic makeup could be modified so that they develop appropriate symptoms.

But this work has been lagging, partly because researchers have had trouble getting pig oocytes to start dividing after the nuclear transfers. Moreover, researchers are still working out a suitable way to keep new embryos alive until they can be placed into female pigs for continued development.

At the meeting, several teams reported progress solving these problems. At the University of Missouri, Columbia, Randall Prather has worked out a new way to activate cell division using a chemical called thimerosal as the initial trigger. And reproductive physiologist Neal First's group at the University of Wisconsin, Madison, offered a more radical potential solution: Avoid the hard-to-activate pig egg altogether by transferring nuclei from adult pigs into bovine oocytes. "Instead of using a pig oocyte, perhaps you could use a sheep or cow oocyte," Robl suggests. It is unclear, however, whether such cross-species embryos would ever come to term.

A High Death Rate Continues with Nuclear Transfer

In spite of the rapid advances in nuclear transfer since Dolly's debut, some big obstacles still remain. At each step along the way some—often many—individuals don't survive. That low efficiency doomed an earlier version of nuclear transfer when it made its commercial debut a decade ago. At that time, several companies, including Granada Inc., based in Houston, were going great guns using nuclei from very early embryos to clone hundreds of calves to make large herds of genetically superior beef cattle. But by 1991, Granada had shut its doors. "We couldn't make as many calves as we wanted to," recalls Bondioli, who worked there. And too often, calves were oversized and unhealthy, with lungs that were not fully developed at birth.

Researchers see the same trends in the few cows and sheep produced by the newer cloning procedures. Large numbers of deaths occur around the time of birth. For example, PPL and Roslin lost eight of II lambs in their first experiment with transgenic clones. But it's not the nuclear transfer procedure itself that's at fault, says Robl. Animals produced by *in vitro* fertilization and other procedures involving the manipulation of embryos have similar problems, albeit at a lower frequency.

"Something that you do to the embryo ... leads to a problem nine months later," says George Seidel Jr., a physiologist at Colorado State University in Fort Collins. His data and other observations suggest that in problem calves the placenta does not function as it should. As a result, cloned calves have too little oxygen and low concentrations of certain growth factors in their blood.

While some researchers are experimenting with different nutrient solutions or making other subtle changes in their nuclear transfer techniques to make embryos and newborns thrive, others are frantically trying to hone the genetic manipulation techniques. Researchers currently have no control over where the foreign genes end up in the chromosomes or how many copies of the gene become part of that cell's genetic repertoire.

Developing that control would enable them to knock out specific genes, say the one encoding the pig protein that elicits a strong, immediate rejection response to pig organ transplants. "The Holy Grail for many is finding a way of getting targeted disruption of genes in livestock as we have in mice," explains Colman, who is confident that even this tough molecular biology problem will be solved quickly. "I expect we'll have targeting solved by next year," he predicts.

Such confidence is required in this fast-moving field, in which progress generally comes through trial and error. Understanding how it all works, say these scientists, will come later. "[There] clearly is at this point in time a pushing forward of the technology," says Alexion's Bondioli. "Have we learned any more biology? Probably not. But [we] have opened up a means to study [it]."

20

The Benefits and Dangers of Cloning Human Cells Are Not Those Most Talked About

J. Madeleine Nash

An elderly man develops macular degeneration, a disease that destroys vision. To bolster his failing eyesight, he receives a transplant of healthy retinal tissue—cloned from his own cells and cultivated in a lab dish.

A baby girl is born free of the gene that causes Tay-Sachs disease, even though both her parents are carriers. The reason? In the embryonic cell from which she was cloned, the flawed gene was replaced with normal DNA. These futuristic scenarios are not now part of the debate over human cloning, but they should be. Spurred by the fear that maverick physicist Richard Seed, or someone like him, will open a cloning clinic, lawmakers are rushing to enact broad restrictions against human cloning. To date, 19 European nations have signed an anticloning treaty. The Clinton Administration backs a proposal that would impose a five-year moratorium. House majority leader Dick Armey has thrown his weight behind a bill that would ban human cloning permanently, and at least 18 states are contemplating legislative action of their own. "This is the right thing to do, at the right time, for the sake of human dignity," said Armey last week. "How can you put a statute of limitations on right and wrong?"

But hasty legislation could easily be too restrictive. Last year, for instance, Florida considered a law that would have barred the cloning of human DNA, a routine procedure in biomedical research. California passed badly worded legislation that temporarily bans not just human cloning but also a procedure that shows promise as a new treatment for infertility.

This appeared as "The Case for Cloning" *Time*, v151 n5, p.81, February 9, 1998.

Most lawmakers are focused on a nightmarish vision in which billionaires and celebrities flood the world with genetic copies of themselves. But scientists say it's unlikely that anyone is going to be churning out limited editions of Michael Jordan or Madeleine Albright. "Oh, it can be done," says Dr. Mark Sauer, chief of reproductive endocrinology at Columbia University's College of Physicians and Surgeons. "It's just that the best people, who could do it, aren't going to be doing it."

Cloning individual human cells, however, is another matter. Biologists are already talking about harnessing for medical purposes the technique that produced the sheep called Dolly. They might, for example, obtain healthy cells from a patient with leukemia or a burn victim and then transfer the nucleus of each cell into an unfertilized egg from which the nucleus has been removed. Coddled in culture dishes, these embryonic clones—each genetically identical to the patient from which the nuclei came—would begin to divide.

The cells would not have to grow into a fetus, however. The addition of powerful growth factors could ensure that the clones develop only into specialized cells and tissue. For the leukemia patient, for example, the cloned cells could provide an infusion of fresh bone marrow, and for the burn victim, grafts of brand-new skin. Unlike cells from an unrelated donor, these cloned cells would incur no danger of rejection; patients would be spared the need to take powerful drugs to suppress the immune system. "Given its potential benefit," says Dr. Robert Winston, a fertility expert at London's Hammersmith Hospital, "I would argue that it would be unethical not to continue this line of research."

There are dangers, but not the ones everyone's talking about, according to Princeton University molecular biologist Lee Silver, author of *Remaking Eden* (Avon Books). Silver believes that cloning is the technology that will finally make it possible to apply genetic engineering to humans. First, parents will want to banish inherited diseases like Tay-Sachs. Then they will try to eliminate predispositions to alcoholism and obesity. In the end, says Silver, they will attempt to augment normal traits like intelligence and athletic prowess.

Cloning could be vital to that process. At present, introducing genes into chromosomes is very much a hit-or-miss proposition. Scientists might achieve the result they intend once in 20 times, making the procedure far too risky to perform on a

eck Out Receipt

L- Egleston Square Branch Library
7-445-4340
tp://www.bpl.org/branches/egleston.htm

ursday, April 28, 2016 5:18:44 PM

em: 39999036677753
tle: Cloning : for and against
terial: Book
: 05/19/2016

al items: 1

nk You!

human embryo. Through cloning, however, scientists could make 20 copies of the embryo they wished to modify, greatly boosting their chance of success.

Perhaps now would be a good time to ask ourselves which we fear more: that cloning will produce multiple copies of crazed despots, as in the film *The Boys from Brazil*; or that it will lead to the society portrayed in *Gattaca*, the recent science-fiction thriller in which genetic enhancement of a privileged few creates a rigid caste structure. By acting sensibly, we might avoid both traps.

21

Unmodified Nature Is Not Always Benign

Virginia I. Postrel

Twenty years ago, the bookstore in which I was working closed for a few hours while we all went to the funeral of one of our colleagues. Herbie was a delightful guy, well liked by everyone. He died in his 20s—a ripe old age back then for someone with cystic fibrosis. In keeping with the family's wishes, we all contributed money in his memory to support research on the disease. In those days, the best hope was that scientists would develop a prenatal test that would identify fetuses likely to have C.F., allowing them to be aborted. The thought made us uncomfortable. "Would you really want Herbie never to be?" said my boss.

But science has a way of surprising us. Two decades later, abortion is no longer the answer proposed for cystic fibrosis. Gene therapy—the kind of audacious high-tech tool that generates countless references to *Brave New World* and *Frankenstein*—promises not to stamp out future Herbies but to cure them.

This spring I thought of Herbie for the first time in years. It was amid the brouhaha over cloning, as bioethicists galore were popping up on TV to demand that scientists justify their unnatural activities and Pat Buchanan was declaring that "mankind's got to control science, not the other way around."

It wasn't the technophobic fulminations of the anti-cloning pundits that brought back Herbie's memory, however. It was a letter from my husband's college roommate and his wife. Their 16-month-old son had been diagnosed with cystic fibrosis. He was doing fine now, they wrote, and they were optimistic about the progress of research on the disease.

There are no Herbies on *Crossfire*, and no babies with

This appeared as "Fatalist Attraction" in *Reason*, v29 (July 1997), pp. 4–6.

deadly diseases. There are only nature and technology, science and society, "ethics" and ambition. Our public debate about biotechnology is loud and impassioned but, most of all, abstract. Cowed by an intellectual culture that treats progress as a myth, widespread choice as an indulgence, and science as the source of atom bombs, even biotech's defenders rarely state their case in stark, personal terms. Its opponents, meanwhile, act as though medical advances are an evil, thrust upon us by scheming scientists. Hence Buchanan talks of "science" as distinct from "mankind" and ubiquitous Boston University bioethicist George Annas declares, "I want to put the burden of proof on scientists to show us why society needs this before society permits them to go ahead and (do) it."

That isn't, however, how medical science works. True, there are research biologists studying life for its own sake. But the advances that get bioethicists exercised spring not from pure science but from consumer demand: "Society" may not ask for them, but individual people do.

Living in a center of medical research, I am always struck by the people who appear on the local news, having just undergone this or that unprecedented medical procedure. They are all so ordinary, so down-to-earth. They are almost always middle-class, traditional families, people with big medical problems that require unusual solutions. They are not the Faustian, hedonistic yuppies you'd imagine from the way the pundits talk.

And it is the ambitions of such ordinary people, with yearnings as old as humanity—for children, for health, for a long and healthy life for their loved ones—of which the experts so profoundly disapprove. As we race toward what Greg Benford aptly calls "the biological century," we will hear plenty of warnings that we should not play God or fool Mother Nature. (See "Biology: 2001," November 1995.) We will hear the natural equated with the good, and fatalism lauded as maturity. That is a sentiment about which both green romantics and pious conservatives agree. And it deserves far more scrutiny than it usually gets.

Nobody wants to stand around and point a finger at this woman (who had a baby at 63) and say, 'You're immoral. But generalize the practice and ask yourself, What does it really mean that we won't accept the life cycle or life course?" Leon Kass, the neocons' favorite bioethicist, told The *New York Times.* "That's one of the big problems of the contemporary scene. You've got all kinds of people who make a living and support themselves but who psychologically are not grown up. We have a culture of functional immaturity."

It sounds so profound, so wise, to denounce "functional immaturity" and set oneself up as a grown-up in a society of brats. But what exactly does it mean in this context? Kass can't possibly think that 63-year-olds will start flocking to fertility clinics—that was the quirky action of one determined woman. He is worried about something far more fundamental: our unwillingness to put up with whatever nature hands out, to accept our fates, to act our ages. "The *good* news," says Annas of human cloning, "is I think finally we have a technology that we can all agree shouldn't be used." (Emphasis added.) Lots of biotech is bad, he implies, but it's so damned hard to get people to admit it.

When confronted with such sentiments, we should remember just what Mother Nature looks like unmodified. Few biotechnophobes are as honest as British philosopher John Gray, who in a 1993 appeal for greens and conservatives to unite, wrote of "macabre high-tech medicine involving organ transplantation" and urged that we treat death as "a friend to be welcomed." Suffering is the human condition, he suggested: We should just lie back and accept it. "For millennia," he said, "people have been born, have suffered pain and illness, and have died, without those occurrences being understood as treatable diseases."

Gray's historical perspective is quite correct. In the good old days, rich men did not need divorce to dump their first wives for trophies. Childbirth and disease did the trick. In traditional societies, divorce, abandonment, annulment, concubinage, and polygamy—not hightech medicine—were the cures for infertility. Until the 20th century, C.F. didn't need a separate diagnosis, since it was just one cause of infant mortality among many. Insulin treatment for diabetes (highly unnatural) didn't exist until the 1920s. My own grandmother saw her father, brother, and youngest sister die before she was in middle age. In 1964 a rubella epidemic left a cohort of American newborns deaf.

These days, we in rich countries have the wonderful luxury of rejecting even relatively minor ailments, from menstrual cramps to migraines, as unnecessary and treatable. "People had always suffered from allergies … But compared to the other health problems people faced before the middle of the twentieth century, the sneezing, itching, and skin eruptions had for the most part been looked at as a nuisance," writes biologist Edward Golub. "In the modern world, however, they became serious impediments to living a full life, and the discovery that a whole class of compounds called antihistamines could control the symptoms of allergy meant that allergic individuals could lead close to normal lives. The same story can be told for high blood pressure, depression, and a large number of chronic conditions."

Treating chronic conditions is, if anything, more nature-defiant than attacking infectious diseases. A woman doesn't have to have a baby when she's 63 to refuse to "accept the life cycle or life course." She can just take estrogen. And, sure enough, there is a steady drumbeat of criticism against such unnatural measures, as there is against such psychologically active drugs as Prozac. We should, say the critics, just take what nature gives us.

In large part, this attitude stems from a naive notion of health as the natural state of the body. In fact, disease and death are natural; the cures are artificial. And as we rocket toward the biological century, we will increasingly realize that a bodily state may not be a "disease," but just something we wish to change. Arceli Keh was not sick because her ovaries no longer generated eggs; she was simply past menopause. To say she should be able to defy her natural clock (while admitting that mid-60s parenthood may not be the world's greatest idea) doesn't mean declaring menopause a disease. Nor does taking estrogen, any more than taking birth control pills means fertility is a sickness.

"The cloned human would be an attack on the dignity and integrity of every single person on this earth," says German Research Minister Juergen Ruettgers, demanding a worldwide ban, lest such subhumans pollute the planet. (The Germans want to outlaw even the cloning of human cells for medical research.) Human cloning is an issue, but it is not the issue in these debates. They are really about whether centralized powers will wrest hold of scientists' freedom of inquiry and patients' freedom to choose—whether one set of experts will decide what is natural and proper for all of us—and whether, in fact, nature should be our standard of value.

Ruettgers is wildly overreacting and, in the process, attacking the humanity of people yet unborn. As Ron Bailey has noted in these pages, human cloning is not that scary, unless you're afraid of identical twins, nor does it pose unprecedented ethical problems. No one has come up with a terribly plausible scenario of when human cloning might occur. Yet judging from the history of other medical technologies, the chances are good that if such a clone were created, the parents involved would be ordinary human being with reasons both quite rare and extremely sympathetic. We should not let the arrogant likes of Ruettgers block their future hopes.

22

Cloning Technology Will Yield Medical Benefits

Robert Winston

The production of a sheep clone, Dolly, from an adult somatic cell[1] is a stunning achievement of British science. It also holds great promise for human medicine. Sadly, the media have sensationalised the implications, ignoring the huge potential of this experiment. Accusations that scientists have been working secretively and without the chance for public debate are invalid. Successful cloning was publicised in 1975,[2] and it is over eight years since Prather et al published details of the first piglet clone after nuclear transfer.[3]

Missing from much of the debate about Dolly is recognition that she is not an identical clone. Part of our genetic material comes from the mitochondria in the cytoplasm of the egg. In Dolly's case only the nuclear DNA was transferred. Moreover, we are a product of our nurture as much as our genetic nature. Monovular twins are genetically closer than are artificially produced clones, and no one could deny that such twins have quite separate identities.

Dolly's birth provokes fascinating questions. How old is she? Her nuclear DNA gives her potentially adult status, but her mitochondria are those of a newborn. Mitochrondia are important in the aging process because aging is related to acquired mutations in mitochondrial DNA, possibly caused by oxygen damage during an individual's life.[4] Experimental nuclear transfer in animals and in human cell lines could help elucidate mechanisms for many of these processes.

Equally extraordinary is the question concerning the role of the egg's cytoplasm in mammalian development. Once the

---

This article appeared as "The Promise of Cloning for Human Medicine" *British Medical Journal,* No 7085 Volume 314, Editorial Saturday 29 March 1997.

quiescent nucleus had been transferred to the recipient egg cell, developmental genes expressed only in very early life were switched on. There are likely to be powerful factors in the cytoplasm of the egg that make this happen. Egg cytoplasm is perhaps the new royal jelly. Studying why and how these genes switch on would give important information about both human development and genetic disease.

Research on nuclear transfer into human eggs has immense clinical value. Here is a model for learning more about somatic cell differentiation. If, in due course, we could influence differentiation to give rise to targeted cell types we might generate many tissues of great value in transplantation. These could include skin and blood cells, and possibly neuronal tissue, for the treatment of injury, for bone marrow transplants for leukaemia, and for degenerative diseases such as Parkinson's disease. One problem to be overcome is the existence of histocompatibility antigens encoded by mitochondrial DNA,[5] but there may be various ways of altering their expression. Cloning techniques might also be useful in developing transgenic animals—for example, for human xenotransplantation.

Cloning Can Help Save Rare Species

There are also environmental advantages in pursuing this technology. Mention has been made of the use of these methods to produce dairy herds and other livestock. This would be of limited value because animals with genetic diversity derived by sexual reproduction will always be preferable to those produced asexually. The risk of a line of farm animals prone to a particular disease would be ever present. However, cloning offers real prospects for preservation of endangered or rare species.

In human reproduction, cloning techniques could offer prospects to sufferers from intractable infertility. At present there is no treatment, for example, for those men who exhibit total germ cell failure. Clearly it is far fetched to believe that we are now able to reproduce the process of meiosis, but it may be possible in future to produce a haploid cell from the male which could be used for fertilisation of female gametes. Even if straight cloning techniques were used, the mother would contribute important constituents—her mitochondrial genes, intrauterine influences, and subsequent nurture.

Regulation of cloning is needed, but British law already

covers this. Talk of "legal loopholes"[6] is wrong. The Human Fertilisation and Embryology Act may need modification, but there is no particular urgency. A precipitate ban on human nuclear transfer would, for example, prevent the use of *in vitro* fertilisation and preimplantation diagnosis for those couples at risk of having children who have appalling mitochondrial diseases.[7] Self regulation and legislation already work well. Apart from any other consideration, it seems highly unlikely that doctors would transfer human clones to the uterus out of simple self interest. Many of the animal clones that have been produced show serious developmental abnormalities,[8] and, apart from ethical considerations, doctors would not run the medicolegal risks involved. Transgenic technology has been with us for 20 years, but no clinician has been foolish enough to experiment with human germ cell therapy. The production of Dolly should not be seen as a moral threat, but rather as an exciting challenge. To answer this good science with a knee jerk political reaction, as did President Clinton recently,[9] shows poor judgment. In a society which is still scientifically illiterate, the onus is on researchers to explain the potential good that can be gained in the laboratory.

References to Reading 22

1 Wilmut T, Schnieke A K, McWhir J, Kind A J, Campbell K H S. Viable offspring derived from fetal and adult mammalian cells. *Nature* 1997;385: 810–3.

2 Gurdon J B, Laskey R A, Reeves O R. The developmental capacity of nuclei transplanted from keratinised skin cells of adult frogs. *J Embryol Exp Morph* 1975;34: 93–112.

3 Prather R S, Simms M M, First N L. Nuclear transplantation in early pig embryos. *Biol Reprod* 1989;41: 414–8.

4 Ozawa T. Mitochondrial DNA mutations associated with aging and degenerative diseases. *Exp Gerontol* 1995;30: 269–90.

5 Dabhi V M, Lindahl K F. MtDNA-encoded histocompatibility antigens. *Methods Enzymol* 1995;260: 466–85.

6 Masood E. Cloning technique "reveals legal loophole." *Nature* 1997;385: 757.

7 Winston R M, Handyside A H. New challenges in human *in vitro* fertilization. *Science* 1993;260: 932–6.

8 Campbell K H S, McWhir J, Ritchie W A, Wilmut I. Sheep cloned by nuclear transfer from a cultured cell line. *Nature*

1996;380: 64–6.
9 Wise J. Sheep cloned from mammary gland cells. *BMJ* 1997;314: 623.

23

A Cloned Gene Yields Stretchy Spider Silk

Mari N. Jensen

To snag a speeding insect, the resilient silk at the center of a spider's web may stretch to almost three times its original length. Now, researchers have cloned the gene for this most elastic of spider silks and unraveled its protein structure.

The extreme elasticity of this natural miracle fiber, called capture silk, comes from long spirals in the protein's configuration, propose researchers from the University of Wyoming in Laramie. Figuring out what makes silk stretchy and what makes it strong will ultimately enable scientists to design genes to control the manufacture of silks, says Randolph V. Lewis, a coauthor of the report in the Feb. 6 *Journal of Molecular Biology*. This finding, he says, "Gives us the tools to say, 'If you want to make an elastic silk, this is what you've gotta have.'"

The researchers obtained the gene from a gland of the golden orb-weaving spider, Nephila clavipes. They found that capture silk protein, a chain of thousands of amino acids, contains regions in which a sequence of five amino acids is repeated over and over, as many as 63 times.

The researchers suggest that the segments of the protein with the repeating blocks form long, springlike shapes. At the end of each five-amino-acid block, the protein kinks back on itself in a 180-degree turn, Lewis says. The series of turns eventually forms a spiral that "looks exactly like a molecular spring."

Spiders make as many as seven different types of silk, says coauthor Cheryl Y. Hayashi. Insects get entangled in the sticky web, she explains, because the stretchiness of capture silk lets the

---

This appeared as "Gene Cloned For Stretchiest Spider Silk" in *Science News*, v153n8, p.119, Feb 21, 1998.

web oscillate back and forth after the insect hits it. If the web were stiff, the insect might just bounce off.

Researchers have cloned several genes for dragline silk, the type that the nursery rhyme spider must have spun to lower itself down beside Miss Muffet. Spiders use dragline silk to form the guylines and framework for wheel-shaped orb webs. It is stronger than capture silk but less flexible (*Science News* 3/9/96, p. 152). In fact, Lewis says, dragline silk is only onefifth as elastic as capture silk.

Dragline silk proteins and capture silk proteins have similar turn-forming blocks of amino acids. However, the researchers found that these blocks repeat an average of 43 times in the capture silk, compared to only nine times in the dragline silk. That fivefold difference in length corresponds to the difference in elasticity between the two proteins, Lewis says.

"When you put the math to it," he notes, "it looks pretty good." The stretchy section of the protein may not spiral in the way Lewis describes, cautions John Gosline, a biomechanic at the University of British Columbia in Vancouver.

An alternative theory suggests that the zigzag turns may simply allow the protein to flex and bend, Gosline says. Rather than assuming a specific, organized shape, the stretchy parts of the protein may flop around at random.

Gosline adds that he has no doubt that the Wyoming group has identified the correct gene for capture silk protein.

"I think it's interesting," he says. "We're actually a bit jealous."

V

The Rights and Wrongs of Cloning

Introduction to Part V

Cloning is mainly discussed in terms of ethical values: good and evil, right and wrong. *National Review* links its objection to cloning to the distinctive uniqueness of human beings (Reading 24). Senator Tom Harkin argues (Reading 25) that it is not wrong to conduct cloning research to help people, and that no area of knowledge should be closed to investigation. By contrast, John Garvey maintains (Reading 26) that application of cloning to human beings endangers our reverence for mystery. A somewhat similar position is taken by *Commonweal* (Reading 27), which also succinctly states a number of ethical objections, including the argument that, "like a gardener who thinks she can manage the ecology of an entire forest," we can never know enough to practice human cloning responsibly.

Finally, two articles which eloquently present more extended arguments, for opposite conclusions. Ronald Bailey (Reading 28) reviews and criticizes the main arguments against allowing human cloning, while Daniel Callahan maintains (Reading 29) that the discipline of bioethics has failed to come to grips with the demonstrated need for a permanent ban on human cloning.

24

Modern Ideologues Have Forgotten What Medievals Knew

The Editors of *National Review*

The first reactions to cloning were surprising in their consensus: Left or Right, there was concern about losing the sense of human beings as unique, as creatures who must be respected in themselves rather than as things that are "made," or manufactured to order, or used as a source of spare parts.

Still, when humanity gets down to cases, the prospects may become more seductive. With the right genetic manipulation, for example, it may be possible to grow, in pigs, livers that will not excite the immune systems of humans when there is a need to transplant these organs. But if much can be gained from a pig, what benefits may be gained from the donations of humans?

G.K. Chesterton remarked that the modern world has reached a condition in which "it is wrong even when it is right." It may do sensible things, but "it is rapidly ceasing to have any of the sensible reasons for doing them." The modern world denounces superstitions, but "its own principal virtues are now almost entirely superstitions." And so we may preserve a revulsion toward cannibalism but people find it harder to explain why. They are more likely to fall back on convention ("That isn't what we do here, or at least not on Fridays"). The matter was explained more readily when people had the sense, with Aristotle, of the critical differences that separated human beings from animals, or when that understanding was fortified by the religious conviction that human beings were made in the image of something higher.

But our own age tends to cast doubt on these convictions and to make much less of the differences between men and animals. As Chesterton observed of cannibalism, "the modern

---

This appeared as "Cloning Cloning Cloning" in *National Review*. v49 (Mar. 24 1997). p. 16.

theorist will have to defend his own sanity with a prejudice. It is the medieval theologian who can defend it with a reason." Left and Right are determined now to draw a bright line between cloning animals and manipulating humans. But what are the prospects for preserving that barrier when, in countless ways, every day, people suggest that they no longer remember, or respect, the reasons behind it?

25

There Are No Appropriate Limits to Human Knowledge

Tom Harkin

[T]here has been a lot of, I think, undue, inflammatory kinds of statements and comments made about this cloning research. It seems odd to me that on something that has so much potential to alleviate human suffering and which is also, I will be frank to admit, fraught with perils of ethics and bioethics—it seems odd to me that a bill of that nature would be rushed so soon to the floor of the Senate.

… each year, too many of our loved ones suffer terribly. They are taken away from us by diseases like cancer, heart disease and Alzheimer's. For many years, I have worked hard to expand research into finding cures and preventative measures and improve treatments for the many conditions that rob us of our health. Over the last several years, there have been major breakthroughs in medical research. We need to make sure that our world-class scientists continue to build on this progress, but that we also say to young people who are in college today, maybe even in high school, who are thinking of pursuing research careers, that we welcome their inquisitiveness, we welcome their experimentation, we want there to be no bounds put on their inquiries by a rush to judgment by the Congress of the United States, which is ill-equipped to make such a judgment. I think our actions here send a very chilling message to young people, who want to go into biomedical research, that somehow there is going to be the heavy hand of "Big Brother" Government overlooking their research, telling them you can do this but not that, or you can go no further than that, or you can ask this question, but you can't ask that question.

Now, another area of research that has been ongoing for a long time—this is nothing new—has recently captured public attention. That is the research into cloning, cloning cells. Now, there is a man in Chicago—I don't know him and I never have met him—and his name is Richard Seed. Well, he caused quite a sensation a few weeks ago by saying he intends to clone infertile people within the

---

This appeared in the *Congressional Record* on February 9, 1998 p S507–8

next 2 years. Well, when I first heard this, I said, who is this guy? I never heard of him and I have been involved in research, medical research for a long time. Well, I found out that, quite frankly, he is a very irresponsible individual. He doesn't have the expertise himself. He doesn't have the laboratory, the money, or the wherewithal. I think most researchers and policymakers that I know who know of this person say that he is both out of the mainstream and that his plans for cloning are, at the very least, premature. Now, again, from all that I have read—and now I have seen him on television—I think that Mr. Seed is more interested in getting his name in the paper than actually carrying out any legitimate scientific research. This is the unfortunate part of it. Why should the irresponsible actions of an individual like Mr. Seed lead to irresponsible actions on our part, because that is exactly what we are doing? Is Mr. Seed irresponsible? I believe so, absolutely. As I said, he doesn't have the expertise, the lab, or the wherewithal to even carry out this research. So he is making very irrational, irresponsible, inflammatory statements. But then why should we respond irresponsibly? I think we should respond responsibly and very carefully to an area of scientific research that can hold so much promise to alleviate pain and suffering and premature death all around the world. Let's not act irresponsibly because one person in America has spoken irresponsibly. ...

... Science makes genetically identical tissues and organs for the treatment of a vast array of diseases. I gave a sort of off-the-cuff set of comments last summer when this issue came up with Dolly, the sheep that was cloned in Scotland. Dr. Wilmut was at our committee. I talked about the need to continue research into cloning of cells. I said it was going to happen in my lifetime. I certainly stand here and hope that it does. Shortly after that, I was at a restaurant in a small town in Iowa. A person came up to me, a friend of mine. I went over to their booth to see them. There was a woman there whom I had never met, a rather young woman with her husband. I was introduced to them. Just right out of the clear blue she said, "Thank you for what you said about cloning and taking the position you did on cloning." I don't even think it was in the newspaper. It was on television, I think. CNN may have carried that type of thing. But I was curious as to why this young woman, who, if I am not mistaken, lives on a farm, I believe—I can't quite remember that detail. I asked her, "Why are you so interested in this?" She said because she has a rare kidney disorder. She is hoping because of rejection possibilities that there might come a time when we could actually grow the kind of tissue that would develop into a kidney to replace her kidney so that there wouldn't be that possibility of rejection. She got it. She understood it. That is what we are talking about. Those are the kinds of possibilities that I believe will happen in my lifetime if we do not act irresponsibly and irrationally.

Last year, during this hearing on human cloning research, someone asked, "Are there appropriate limits to human knowledge?" Quite frankly, I responded—and I respond again—to say that I do not think there are any appropriate limits to human knowledge, none whatsoever. I think it is the very essence of our humanity and human nature. As long as science is done ethically and openly and with the informed consent of all parties, I do not think Congress should attempt to place limits on the pursuit of knowledge. To those who suggest that cloning research is an attempt to play God, I invite you to take your ranks alongside Pope Paul V who, in 1616, persecuted the great astronomer Galileo for heresy—for saying that the Earth indeed revolved around the Sun and not otherwise.

But we don't have to go back that far. Not too long ago in our Nation's history, Americans viewed artificial insemination as abhorrent and its use was banned as being morally repugnant—even for animals; even for animals. There was an attempt to ban artificial insemination. Of course, now that is about all we use on the farm these days. Heart transplants were scorned and X-rays were considered witchcraft. But today we don't think twice about test tube babies, *in vitro* fertilization, or organ transplants.

Throughout the 1950s, whenever we pushed the bounds of human knowledge, there has always been a constant refrain of saying, "Stop—you are playing God." But if a couple did not have a baby and decides to seek artificial insemination, is that playing God? If a patient is dying of kidney disease and a doctor decided to transplant healthy kidneys, is that playing God? If a patient is dying of heart disease and receives a heart transplant, are we playing God?

Others say that human cloning research is demeaning to human nature. I am sorry; I don't think so. I think that any attempt to limit the pursuit of human knowledge is demeaning to human nature. I think it is the very essence of our humanity to ask how and why and if and what. I think it is demeaning to human nature to raise unfounded fears among the people of America. I think that is demeaning to human nature.

As I said, I think the finest part and the very essence of our human nature and our humanity is to ask why, how, and what if. It is our very humanity that compels us to probe the universe from the subatomic to the cosmos, and, yes, from blastocysts to the full human anatomy. Our humanity compels us to do that.

... What we are talking about here is not the cloning of a human being. What we are talking about is the cloning of cells, and without further research and appropriate regulations, many people will die and become ill and spend very, very miserable lives when that could otherwise be alleviated through this cloning research.

Right and wrong? It is wrong to conduct cloning research that might enable us to grow a liver out of a person's own DNA? To grow skin out of a person's own DNA? Perhaps even to grow heart tissue, or even a full heart, out of a person's own DNA, so there would be no rejection possibilities? It is wrong to do research in cloning of cells that might permit my nephew, Kelly, who, at the age of 19, got injured in the military, his spinal cord was broken and he has been a quadriplegic since and still holds out the hope that research someday is going to enable him to walk again? And, yes, cloning research might be able to rebuild those kinds of cells from his own DNA that will get those nerve endings going again so that my nephew can walk again. That research is wrong?

... Dr. Seed from Chicago is not going to clone any human being. No reputable scientist or doctor that I have spoken to, and I have spoken to quite a few of them, believes he is anywhere near that for years and years and years. But he is making a name for himself. He is on all the talk shows, that's for sure. He has become notorious, a public figure, and I guess a lot of people like to do that. But just because he's irresponsible doesn't mean we ought to be irresponsible.

26

We Should Not Do Everything We Can Do

John Garvey

Years ago Gore Vidal said that he thought human beings had already been cloned. "Anyone who says 'Have a nice day is a clone," he said. "There's a big loaf of them and they just keep slicing them off."

When a Scottish researcher cloned a sheep recently, the event set off another round of joking, some of it quite funny; it also led, predictably, to serious comments by ethicists, theologians, and scientists. Almost everyone seems to assume that human beings will one day be cloned, and that the uses to which cloned humans might be put range from the nightmarish to the ludicrous. People could be grown for body parts—a liver with your exact DNA would probably not need an anti-rejection drug if it were transplanted. One startling suggestion was that a dying child could be cloned, as a sort of replacement kid. There were other ethical problems: a number of fertilized embryos were destroyed along the way toward cloning the sheep.

In an article in the *New York Times* Gustav Niebuhr quoted Notre Dame's Richard McCormick, who said that the obvious motives for cloning a human were "the very reasons you should not." To attempt to create people with specific characteristics is to make single or multiple aspects of being human more important than the "beautiful whole that is the human person." This gets it about right, I think. A more alarming sentiment was expressed by Nancey Murphy of the Fuller Theological Seminary in Pasadena. She said she hoped that ethicists "would concentrate their efforts on saying what we should do with this, rather than saying it should't be done, because people have rightly said it can't be prevented."

How's that? If something can be done and will be done we

This article appeared as "The Mystery Remains" *Commonweal.* v124 (March 28 1997), pp. 6–7.

shouldn't say that it shouldn't be done? Apply that to genocide or euthanasia. Or for that matter the torture of small children. Knowing that adultery and murder will happen does not lead us to ask how they can best be used; we have to say these things shouldn't be done at all. The role of the ethicist is not to refrain from such judgments, but very often to make them.

Some things are clear: A cloned human being is still a human being, with all sorts of individual complications. In essence, cloning is the artificial fashioning of an identical twin, one that will be younger than its sibling. As in the case of identical twins, the similarities will be intriguing, but so will the inevitable differences.

We Don't Own Our Bodies

The real danger is the way the process moves us ever closer to the idea of the human being as product or property. Prisoners in China have been used as involuntary organ donors. A doctor I know has been asked more than once to determine the sex of an unborn child; if the child turned out to be a female, the parents planned to abort it. There is something Nazi-like about wanting to create "desirable" human beings, but we already use people as if they were property (as in the case of the prisoners) or try to create desirable children (as in the case of people who prefer male to female babies). The argument that we have a right to choose our time of death or to abort a child is similar: "Whose life, or whose body, is it anyway?" Our answer must be: not yours, not mine, and not the state's. The body is not a form of property at all, except to a slavemaster.

There are ethicists who have no problem with the idea of cloning itself, but who worry only about its potential for misuse. This strikes me as being tone-deaf to mystery, to any sense of the sacred. The loss of this sense is almost a hallmark of our age, and in some sense the churches have been complicit in it. When churchgoers cannot find any sense of the sacred or of mystery in liturgy, where will they begin to learn to see it in natural processes or the physical world? "Mystery" has taken on a muddled meaning: it is either the "booga booga" stuff that hovers around such TV shows as *The X Files,* or it is seen as something so far beyond understanding that we might as well not think about it at all.

To Countenance Cloning Is to Lose the Sense of Mystery

But the sense of mystery and glory to be found in the best music (Bach or Arvo Pärt) and in the stillness at the heart of the vision of some wonderful painters and iconographers (Giotto or Andrei Rublev) can be experienced in the presence of natural phenomena, and should be. A reverence for mystery does not put a stop to our understanding or our desire to understand; rather, this desire is deepened. The desire to know is experienced in a context marked by deep humility and gratitude. A sense of the sacred, the ability to be still before the mystery, is something which locates us exactly: we are where we are meant to be, who we are meant to be, in such moments.

When science becomes exclusively reductionist, it loses this sense. But it doesn't need to. Retaining this sense of wonder could lead a scientist to say, "Yes, we can do this; it would be interesting, and wrong." The problem here is partly the confusion of science with technology. That something can be done obviously does not mean that it should be; but without a restoration of the sense of the sacred, of mystery, we will probably not be able to begin to make that argument, or even to understand it ourselves.

27

Cloning Humans is Inhuman

Commonweal

From Scotland comes news that a team of fearless scientists has cloned a sheep. One might consider this development one worrisome step for sheep, but more likely it is a very big and very dangerous step for humankind.

"Dolly," as she was christened, was produced in a laboratory where a cell taken from the udder of one sheep was fused with another sheep's egg, from which the nucleus had been removed. The resulting embryo was then implanted in a surrogate mother and brought to term. Dolly is the genetic twin of the sheep from whom the cell was first taken, and her arrival promises benefits in scientific knowledge, agriculture, and medicine.

Dolly caused an uproar in large part because, though theoretically possible, it had long proved technically unfeasible to clone mammals. It is now expected that it is only a matter of time before someone succeeds in cloning the most successful mammal of all—namely, humans. At this point in the conquest of nature by science, it is important to reassert that not everything that can be done should be done. Even Ian Wilmut, the scientist chiefly responsible for bringing Dolly into the world, considers the idea of cloning human beings "offensive ..., ethically unacceptable." His instincts are sound. However, a critical observer might also ask why someone opposed to the likely uses of this technology nevertheless decided to set us off on this path.

Some embrace the prospect of manufacturing human life in laboratories. Such thinking seems more a confirmation of the modern trivialization of the meaning of sex than any sober assessment of what is at stake when technology plays so large a part in human reproduction. The Catholic position on these questions is clear-cut—perhaps too clear-cut. Because it draws the

---

This appeared as "Cloning Isn't Sexy" in *Commonweal*, v124 (Mar. 28 1997). pp. 5–6.

line on intervention in procreation at contraception, the church's often astute warnings about the dehumanizing of sex and reproduction have fallen on deaf ears. That's unfortunate, because the human values the church rightly defends in questioning advanced reproductive technologies make its hair-splitting over "barrier methods" and the so-called "contraceptive mentality" seem like mere intellectualized prudery. In reality, much more is at stake.

There are sound moral reasons why human communities have always tied sexual desire to love, and love to marriage, and marriage to the care of children. Neither the extreme view demanding that every sexual act be "open" to procreation, nor the modern presumption that we should be free to desexualize and depersonalize the act of procreation is the best way to promote human flourishing. But where to draw the line along the continuum of interventions is difficult. In the acceptance of each new technology—artificial insemination, *in vitro* fertilization, surrogate motherhood—a logic of justification is advanced that makes the next moral hurdle seem lower still. Yet as Dr. Wilmut's own trepidation attests, there is widespread uneasiness over giving scientists and potential DNA donors the ability to determine the entire genetic make-up of new human lives. Cloning, with the genetic manipulation and engineering it heralds, may be a line even many who champion "reproductive freedom" will not want to cross.

Still, it is argued that clones ought to be considered little more than "delayed twins." That is true, in a strictly genetic sense. Certainly identical twins occur in nature. But why should anyone be allowed to determine the entire genetic identity of another person? Granting such power over someone else's life—even the life of one's own offspring—is an unwarranted circumscribing of individuality and human possibility. We simply don't have the right to decide such things for others. Parents are entrusted with the lives of their children; they are not the owners or determiners of those lives. As Daniel Callahan has written (*New York Times*, February 26, 1997), cloning "would be a profound threat to what might be called the right to our own identity."

Human cloning would play havoc with notions of parenthood, kinship, the distinct dignity of children in regard to their parents, and perhaps even the sanctity of life. The possibilities for mischief seem endless, and endlessly dizzying. A woman could, for example, give birth to her own twin. As with justification for most reproductive technologies, the desire of an infertile couple for a genetic child may prove to be the most compelling reason for resorting to cloning. Infertility can be a terrible personal loss. Still, infertility is not sufficient justification for any and all means of bringing a child into the world. Cloning would represent yet a further commodification of procreation; in its asexual method of reproduction and the genetic asymmetry of the child produced, cloning further relativizes the most fundamental of human relationships: that of wife and husband and parents and children. Technological wizardry must not be allowed to undermine monogamous marriage and the biological family. Science and technology should serve the common good, not its own self-aggrandizing imperatives or mere individual desire.

When it comes to cloning, we are a bit like a gardener who thinks she can manage the ecology of an entire forest. There is too much we do not know, too much we can never know. Worse, in almost every respect—from the discarding of "surplus" embryos to the possible creation of genetic monstrosities—cloning requires that we regard another human being as a means to an end, and not as an end in itself. Indeed, the very attempt to clone humans constitutes experimentation (it is not a medical procedure) on someone who is incapable of giving consent, a violation of the most fundamental principles of medicine and science. In sum, like abortion and euthanasia, cloning human beings moves us a step closer to an openly utilitarian definition of human dignity and life.

Justified in the name of scientific progress and humanitarian relief, cloning will be a powerful temptation. Yet the intuition that tells us there is something inherently inhuman in the laboratory production of human beings is sound. Science and technology enable us to transcend our physical limitations, but in doing so we run the risk at a certain point of betraying our true natures. In the creation of children it is important that we prize our full, embodied humanity, and not just one aspect of it. When it comes to separating reproduction from human sexuality, the siren call of scientific progress is largely a ruse. The kind of control over human life and destiny promised by these new reproductive technologies is far too potent to be left in the hands of scientists or anyone else. As our brave new world of self-creation unfolds, we must guard against progress in the name of humanity that in fact dehumanizes.

28

The Standard Objections to Cloning Won't Bear Examination

Ronald Bailey

By now everyone knows that Scottish biotechnologists have cloned a sheep. They took a cell from a six-year-old sheep, added its genes to a hollowed-out egg from another sheep, and placed it in the womb of yet another sheep, resulting in the birth of an identical twin sheep that is six years younger than its sister. This event was quickly followed up by the announcement that some Oregon scientists had cloned monkeys. The researchers say that in principle it should be possible to clone humans. That prospect has apparently frightened a lot of people, and quite a few of them are calling for regulators to ban cloning since we cannot predict what the consequences of it will be.

President Clinton rushed to ban federal funding of human cloning research and asked privately funded researchers to stop such research at least until the National Bioethics Advisory Commission issues a report on the ethical implications of human cloning. The commission, composed of scientists, lawyers, and ethicists, was appointed last year to advise the federal government on the ethical questions posed by biotechnology research and new medical therapies ...

But Senator Christopher Bond (Republican–Missouri) isn't waiting around for the commission's recommendations; he's already made up his mind. Bond introduced a bill to ban the federal funding of human cloning or human cloning research. "I want to send a clear signal," said the senator, "that this is something we cannot and should not tolerate. This type of research on humans is morally reprehensible."

Carl Feldbaum, president of the Biotechnology Industry

This appeared as "The Twin Paradox: What Exactly Is Wrong with Cloning People?" *Reason*, v29 (May 1997). pp. 52–4.

Organization, hurriedly said that human cloning should be immediately banned. Perennial Luddite Jeremy Rifkin grandly pronounced that cloning "throws every convention, every historical tradition, up for grabs." At the putative opposite end of the political spectrum, conservative columnist George Will chimed in: "What if the great given—a human being is a product of the union of a man and woman—is no longer a given?"

In addition to these pundits and politicians, a whole raft of bioethicists declared that they, too, oppose human cloning. Daniel Callahan of the Hastings Center said flat out: "The message must be simple and decisive: The human species doesn't need cloning." George Annas of Boston University agreed: "Most people who have thought about this believe it is not a reasonable use and should not be allowed ... This is not a case of scientific freedom vs. the regulators."

Clones are People

Given all of the brouhaha, you'd think it was crystal clear why cloning humans is unethical. But what exactly is wrong with it? Which ethical principle does cloning violate? Stealing? Lying? Coveting? Murdering? What? Most of the arguments against cloning amount to little more than a reformulation of the old familiar refrain of Luddites everywhere: "If God had meant for man to fly, he would have given us wings. And if God had meant for man to clone, he would have given us spores." Ethical reasoning requires more than that.

What would a clone be? Well, he or she would be a complete human being who happens to share the same genes with another person. Today, we call such people identical twins. To my knowledge no one has argued that twins are immoral. Of course, cloned twins would not be the same age. But it is hard to see why this age difference might present an ethical problem—or give clones a different moral status.

"You should treat all clones like you would treat all monozygous (identical) twins or triplets," concludes Dr. H. Tristam Engelhardt, a professor of medicine at Baylor and a philosopher at Rice University. "That's it." It would be unethical to treat a human clone as anything other than a human being. If this principle is observed, he argues, all the other "ethical" problems for a secular society essentially disappear. John Fletcher, a professor of biomedical ethics in the medical school at the

University of Virginia, agrees: "I don't believe that there is any intrinsic reason why cloning should not be done."

Let's take a look at a few of the scenarios that opponents of human cloning have sketched out. Some argue that clones would undermine the uniqueness of each human being. "Can individuality, identity and dignity be severed from genetic distinctiveness, and from belief in a person's open future?" asks George Will.

Will and others have apparently fallen under the sway of what Fletcher calls "genetic essentialism." Fletcher says polls indicate that some 30 percent to 40 percent of Americans are genetic essentialists, who believe that genes almost completely determine who a person is. But a person who is a clone would live in a very different world from that of his genetic predecessor. With greatly divergent experiences, their brains would be wired differently. After all, even twins who grow up together are separate people—distinct individuals with different personalities and certainly no lack of Will's "individuality, identity and dignity".

In addition, a clone that grew from one person's DNA inserted in another person's host egg would pick up "maternal factors" from the proteins in that egg, altering its development. Physiological differences between the womb of the original and host mothers could also affect the clone's development. In no sense, therefore, would or could a clone be a "carbon copy" of his or her predecessor.

What about a rich jerk who is so narcissistic that he wants to clone himself so that he can give all his wealth to himself? First, he will fail. His clone is simply not the same person that he is. The clone may be a jerk too, but he will be his own individual jerk. Nor is Jerk Sr.'s action unprecedented. Today, rich people, and regular people too, make an effort to pass along some wealth to their children when they die. People will their estates to their children not only because they are connected by bonds of love but also because they have genetic ties. The principle is no different for clones.

Senator Bond and others worry about a gory scenario in which clones would be created to provide spare parts, such as organs that would not be rejected by the predecessor's immune system. "The creation of a human being should not be for spare parts or as a replacement," says Bond. I agree. The simple response to this scenario is: Clones are people. You must treat them like people. We don't forcibly take organs from one twin and give them to the other. Why would we do that in the case of clones?

The technology of cloning may well allow biotechnologists to develop animals which will grow human-compatible organs for transplant. Cloning is likely to be first used to create animals that produce valuable therapeutic hormones, enzymes, and proteins.

Cloning Does Not Seriously Threaten Diversity

But what about cloning exceptional human beings? George Will put it this way: "Suppose a cloned Michael Jordan, age 8, preferred violin to basketball? Is it imaginable? If so, would it be tolerable to the cloner?" Yes, it is imaginable, and the cloner would just have to put up with violin recitals. Kids are not commercial property—slavery was abolished some time ago. We all know about Little League fathers and stage mothers who push their kids, but given the stubborn nature of individuals, those parents rarely manage to make kids stick forever to something they hate. A ban on cloning wouldn't abolish pushy parents.

One putatively scientific argument against cloning has been raised. As a National Public Radio commentator who opposes cloning quipped, "Diversity isn't just politically correct, it's good science." Sexual reproduction seems to have evolved for the purpose of staying ahead of evermutating pathogens in a continuing arms race. Novel combinations of genes created through sexual reproduction help immune systems devise defenses against rapidly evolving germs, viruses, and parasites. The argument against cloning says that if enough human beings were cloned, pathogens would likely adapt and begin to get the upper hand, causing widespread disease. The analogy often cited is what happens when a lot of farmers all adopt the same corn hybrid. If the hybrid is highly susceptible to a particular bug, then the crop fails.

That warning may have some validity for cloned livestock, which may well have to live in environments protected from infectious disease. But it is unlikely that there will be millions of clones of one person. Genomic diversity would still be the rule for humanity. There might be more identical twins, triplets, etc., but unless there are millions of clones of one person, raging epidemics sweeping through hordes of human beings with identical genomes seem very unlikely.

But even if someday millions of clones of one person existed, who is to say that novel technologies wouldn't by then be able to control human pathogens? After all, it wasn't genetic diversity that caused typhoid, typhus, polio, or measles to all but disappear in the United States. It was modern sanitation and modern medicine.

There's no reason to think that a law against cloning would make much difference anyway. "It's such a simple technology, it won't be ban-able," says Engelhardt. "That's why God made offshore islands, so that anybody who wants to do it can have it done." Cloning would simply go underground and be practiced without legal oversight. This means that people who turned to cloning would not have recourse to the law to enforce contracts, ensure proper standards, and hold practitioners liable for malpractice.

There's No Consensus on the Ethics of Cloning

Who is likely to be making the decisions about whether human cloning should be banned? When President Clinton appointed the National Bioethics Advisory Commission last year, his stated hope was that such a commission could come up with some sort of societal consensus about what we should do with cloning.

The problem with achieving and imposing such a consensus is that Americans live in a large number of disparate moral communities. "If you call up the Pope in Rome, do you think he'll hesitate?" asks Engelhardt. "He'll say, 'No, that's not the way that Christians reproduce. And if you live Christianity of a Roman Catholic sort, that'll be a good enough answer. And if you're fully secular, it won't be a relevant answer at all. And if you're in-between, you'll feel kind of generally guilty."

Engelhardt questions the efficacy of such commissions: "Understand why all such commissions are frauds. Imagine a commission that really represented our political and moral diversity. It would have as its members Jesse Jackson, Jesse Helms, Mother Teresa, Bella Abzug, Phyllis Schafly. And they would all talk together, and they would never agree on anything ... Presidents and Congresses rig—manufacture fraudulently—a consensus by choosing people to serve on such commissions who already more or less agree ... Commissions are created to manufacture the fraudulent view that we have a consensus."

Unlike Engelhardt, Fletcher believes that the National Bioethics Advisory Commission can be useful, but he acknowledges that "all of the commissions in the past have made recommendations that have had their effects in federal regulations. So they are a source eventually of regulations." The bioethics field is littered with ill-advised bans, starting in the mid-1970s with the two-year moratorium on recombining DNA and including the law against selling organs and blood and Clinton's recent prohibition on using human embryos in federally funded medical research.

Humans Must Be Free to Experiment

As history shows, many bioethicists succumb to the thrill of exercising power by saying no. Simply leaving people free to make their own mistakes will get a bioethicist no perks, no conferences, and no power. Bioethicists aren't the ones suffering, the ones dying, and the ones who are infertile, so they do not bear the consequences of their bans. There certainly is a role for bioethicists as advisers, explaining to individuals what the ramifications of their decisions might be. But bioethicists should have no ability to stop individuals from making their own decisions, once they feel that they have enough information.

Ultimately, biotechnology is no different from any other technology—humans must be allowed to experiment with it in order to find its best uses and, yes, to make and learn from mistakes in using it. Trying to decide in advance how a technology should be used is futile. The smartest commission ever assembled simply doesn't have the creativity of millions of human beings trying to live the best lives that they can by trying out and developing new technologies.

So why is the impulse to ban cloning so strong? "We haven't gotten over the nostalgia for the Inquisition," concludes Engelhardt. "We are people who are post-modernist with a nostalgia for the Middle Ages. We still want the state to have the power of the Inquisition to enforce good public morals on everyone, whether they want it or not."

29

The Cloning Issue Reveals Shortcomings in Bioethics

Daniel Callahan

The present debate on cloning should by now have made perfectly clear an enormous shortcoming in bioethics. As a field—in its available methods—it simply has few helpful tactics, insights, or even good provisional strategies, to respond to novel biological developments. They are treated as if the usual ways of dealing with moral problems were adequate, as readily adaptable to something new as to something old. Cloning was first debated in the early to mid-1970s, and the latest Dolly-inspired round has added little new to what was said at that time[1]. Whether this is because of a lack of good theory, or the long quiescent time that passed before Dolly arrived, or an inability to transcend ideological or other constraints is not clear; probably all of them together.

I thought about all this, in a dispirited way, as I read the findings of the National Bioethics Advisory Commission, *Cloning Human Beings*. I was further nudged to think about it when I heard the report, in mid-July, that a Japanese researcher believes he has now perfected the prototype of an artificial womb. When and if that device actually arrives, will we once again be treated to the by-now familiar ritual: the forming of a national commission, the hearing of witnesses and the commissioning of papers, the issuing of a report—and the taking of a moderate middle-of-the-road regulatory stance? Probably so. No better approach is on the horizon.

But we might begin working toward one by looking carefully at the commission's report as an example of the problems the field faces. The report's shortcomings reflect more those of bioethics than the work of the commission.

A word first about the report. It is serious and

This appeared as "Cloning: The Work Not Done" in *The Hastings Center Report,* v27 (September/October 1997), pp. 18–20.

conscientious, balanced in its judgments, and fair in its deference to competing viewpoints. Working under conditions of unusual haste, it did about as good a job in sorting through the issues as one could hope for from such a commission.

Would more time have led to a better report? Not necessarily. Like bioethics itself, the report is stronger procedurally than substantively. The commission effectively melded into a set of coherent (though eminently challengeable) conclusions the full range of conventional perspectives on novel biological developments. It caught well both the sharpest extremes heard in testimony—from apocalypse now if human cloning is accepted to a what's-the-big-deal-anyway shrug—and the range of possibilities in between. Its conclusions were hardly radical, but that is not to be expected of a commission.

Given the state of public opinion, highly antagonistic toward human cloning, and—far more important—the lack of strong support at present within the scientific community for research in that direction, the commission's call for a legal ban on research was hardly a surprise. Nor was it surprising that it wanted a sunset clause of five years on such a ban. That two-part move caught the spirit of the moment—against human cloning—without doing that which is taboo—permanently prohibiting scientific research. The idea of a sunset clause was the perfect *via media,* of a kind that commissions traditionally seek when opinion is radically divided. In that respect, it was a good political solution, attempting to balance a variety of values and interests.

But its political strengths betray its ethical weaknesses. There are three in particular worth noting: (1) the thin moral reasons given for recommending a sunset clause; (2) the dearth of any serious discussion about what the children of the future need for a good life; and (3) the absence of a public interest or common good perspective on the appropriate limits of scientific research.

The Rationale For A Sunset Provision

The rationale for a sunset provision is this: "As scientific information accumulates and public discussion continues, a new judgment may develop and we, as a society, need to retain the flexibility to adjust our course in this manner." This is a disturbing claim. Either it means to apply to the human cloning issue only or to ethical judgments in general.

If human cloning only is meant, the commission should

have indicated what kind of "scientific information" would or could lead to a different ethical conclusion. Did the commission have in mind only the safety issue, or what? And since Dr. Wilmut's technique, while scientifically new, has excited no new moral issues or arguments, it is hard to imagine some decisively fresh insights or change of public mood appearing so soon on the scene. Since the commission obviously thought otherwise, and since the idea of a sunset clause was one of its most important conclusions, it owed us a more detailed account of what it had in mind.

It is possible, however, that the implicit premise of the rationale for a sunset provision is that all moral judgments should be open to new information and the fruits of public discussion. If so, then should it not apply also to the right of informed consent, reproductive choice, and the reprehensibility of slavery? Should the Nuremberg Code, for instance, now just fifty years old, have had a sunset clause? If not, it is hardly evident why such a provision is more pertinent to cloning.

Observation I: Bioethics offers few helpful insights into the way (a) scientific information should affect moral appraisal, or (b) why some judgments should be open to revision but others not.

The Needs Of Children

It is a remarkable testimony to the power of cognitive dissonance that there is nowhere in the report any reflection on the needs of the coming generations of children and how cloning, or biomedical research more broadly, might respond to those needs. There are, to be sure, many useful warnings about potential hazards to individual children. But that is not the same as reflection on the general needs of children. There are also plentiful references to reproductive rights, but for the most part it is the welfare and desires of would-be parents, not the needs of children, that are at the core of that notion. The emphasis on safety is the only point of moral certitude expressed in the report. But the need to spare would-be children from experimental danger is not the same as expressing some ideals about the procreation of children, and it is hard to find much on that subject.

Nowhere has anyone suggested that cloning would advance the cause of children. And why should anyone? Apart from imaginative exercises in which cloning would answer some parental wishes, or maybe (in the most idiosyncratic case) save a child's life, children in our world do not suffer from an absence of cloning. But it has been one of the enduring failures of the reproductive rights movement that it has, in the pursuit of parental discretion and relief of infertility, constantly dissociated the needs of children and the desires of would-be parents. Instead of taking the high road and focusing on what children require for a good life (only part of which is being wanted by their parents), the reproductive rights movement consistently drifts toward a lower standard.

What is that standard? It can be seen in the following widespread assumption: since there are now no social constraints upon procreation, and since all sorts of unwise and unregulated procreation already takes place, it is somehow wrong to uphold any procreative ideals for society. This assumption leads to a kind of lazy moral logic: if it is within the rights of even the most irresponsible people to procreate, it surely should be within the rights of conscientious parents who might have their own reasons for wanting a cloned child. In other words, if anything now goes, then it would be an offense to procreative rights not to extend that permissive sanction to human cloning. If anything goes, then anything goes.

I do not want to suggest that the commission took that line. But by virtue of its failure to even ask about the ideals our society should pursue as new means of procreation appear, it showed itself an unwitting captive of two reigning ideologies: that it is wrong to link the pursuit of reproductive rights to the pursuit of the welfare of children in general; and that, in a pluralistic society, it is not legitimate to seek agreement on moral ideals, even about how to procreate and raise children.

The commission might respond to this complaint by noting that it simply had no time to take up such weighty and important matters. Up to a point that is a fair response. But it did have time to take up the safety issue, not necessarily a more important matter (though surely less controversial).

Observation II: When it comes to policy matters, bioethics ordinarily embraces a minimalist ethic, fearful to devise or argue for community ideals or to challenge the individualism of the reproductive rights ideology. It goes with the cultural tide, rarely challenging it.

The Limits of Scientific Research

The commission showed exemplary restraint in its refusal to accept uncritically claims of scientific benefit in human cloning and in its recognition of appropriate moral limits on scientific research. Yet because the issue of safety was the only one that seemed to command a consensus, it left up in the air how a ban could be sustained if (a) the safety problem in human cloning research could be solved, and (b) greatly increased individual and scientific benefits from advances in cloning could be projected. Indeed, in the conditions it offered for limits on inquiry, the stage is well set to overthrow the presently proposed ban: "that the limits are not arbitrary, that they emerge from a thoughtful balancing of costs and benefits, that they are not necessarily oppressive ... "

But we should well know from experience that standards of that kind, which put the burden of proof on those who would limit inquiry, are all too easily met when there is strong scientific pressure to go forward with the research and when aggressive lay groups contend that a ban on inquiry harms their needs and interests. Bans and moratoriums have come and gone. The only exception to that generalization I can think of is the firmness with which the standard of informed consent for human subject research in medicine has been upheld for many years now. But it has been upheld precisely because it has not been allowed to be weighed on a cost-benefit balancing scale, and because it has been allowed to be "oppressive" to some researchers and potential research beneficiaries.

I wonder, for that matter, if the absolute nature of the present requirement of informed consent could be upheld if it was only now being proposed for the first time. Could it these days run the gauntlet of demands for pursuing research possibilities that could relieve infertility, or find a cure for cancer, or provide new basic knowledge (to cite the now-ritualized list of goodies that unfettered research would supposedly bring)? I doubt it. If this seems a dark speculation, take a look at the status of informed consent to experimentation in social psychology, where a deliberate deception of research subjects is widely accepted and even more widely practiced. But then that field—not covered by the Nuremberg or other international codes—uses a standard of balancing benefits and burdens to make its ethical judgments; and, guess what, the benefits of the research usually seem, to the researchers, to outweigh the burden on subjects.[2]

If research on human cloning—which only a handful of scientists have contended is a necessary and needed area for exploration—can now command no more than a five-year ban, it is hard to see how it could fail to go ahead in the future if some distinct general, and not just idiosyncratic, benefits could be discerned. The moral and social scales in favor of research and the relief of illness are already heavily weighted in that direction. It will not take much to tip them in this case.

Observation III: Bioethics has badly neglected the aims and aspirations of research as an area worthy of moral exploration. It has not taken on, for careful examination, the implicit models of human life and welfare and the human future that lie behind the biomedical research enterprise. If it has not been utterly captured by that enterprise, it has mainly stood on the sidelines, wagging its finger now and then. That is no longer good enough.

Footnotes to Reading 29

1. See, for example, Paul Ramsey, *Fabricated Man: The Ethics of Genetic Control* (New Haven: Yale University Press, 1970), especially Chapter 2; Leon R. Kass, "Making Babies—The New Biology and the 'Old Morality'," *The Public Interest*, no. 26 (1972), pp. 18–56; Hans Jonas, *Philosophical Essays* (Englewood Cliffs, N.J.: Prentice-Hall. 1974), pp. 153–63; Francis C. Pizzulli, "Asexual Reproduction and Genetic Engineering: A Constitutional Assessment of the Technology of Cloning," *Southern California Law Review* 47 (February 1974), pp. 476–584.

2. See, for instance, Andreas Ortmann and Ralph Hertwig, "Is Deception Acceptable?" *American Psychologist* 52 (July 1997), pp. 746–47, for a useful account of the status of informed consent in social psychology research.

VI

Cloning and Religion

Introduction to Part VI

In Part VI we present divergent views on the religious implications of human cloning. A report from the *Christian Century* (Reading 30) shows that not all religious leaders have supported the complete prohibition of cloning. John Frederic Kilner (Reading 31) offers several religiously-based objections to cloning of humans, including the harmfulness of separating procreation from marriage. Jean Bethke Elshtain (Reading 32) argues against "the rule of sovereign selves" and suggests that cloning might "turn children into objects and products." Stephen Garrard Post, in Reading 33, submits numerous arguments against cloning but maintains the most important one is that it shows disrespect for nature and nature's God. Finally, in Reading 34 Gregory Pence offers rebuttals to some frequently mentioned grounds for religious opposition to cloning.

30

Some Religious Leaders Would Permit Cloning of Humans

The Christian Century

Theologians, medical ethicists and public policy leaders around the world were among those who took sharp interest in news that a Scottish scientist had successfully cloned a sheep from the mammary cell of a ewe, creating a genetically identical animal without benefit of a male parent. For some the successful experiment posed no threat to the common good or religious morality. Others viewed Ian Wilmut's livestock research at Scotland's Roslin Institute as a further step in an unacceptably high-risk enterprise which meddles with God's creative work.

On March 4 President Clinton, warning scientists against "trying to play God," issued an executive order banning the use of federal funds to pay for human cloning research. "Each human life is unique, born of a miracle that reaches beyond laboratory science," Clinton said. "I believe we must respect this profound gift and resist the temptation to replicate ourselves." The World Health Organization, an agency related to the United Nations, has also stated that attempts to clone human beings should be banned. "WHO considers the use of cloning for the replication of human individuals to be ethically unacceptable," Hiroshi Nakajima, director-general of the UN agency, declared in a statement released March 11.

Vatican officials called for an outright ban on all human cloning and urged scientists not to genetically alter animal species. Southern Baptist and United Methodist leaders in the U.S. issued similar calls for a cloning ban. But other religious thinkers resisted the idea of automatically outlawing new genetic discoveries, suggesting instead a moratorium on cloning experiments until

---

This appeared as "To Clone or Not To Clone?" in *The Christian Century*. v114 (Mar. 19–26 1997). pp. 286–8.

scientists and ethicists sort out the issues.

Some Christians Call for Caution in Condemning Cloning

"One of the most important things people of faith must do is to get their facts straight," said Robert Russell, executive director of the Center for Theology and the Natural Sciences in Berkeley, California. "You can't just take a (religious) tradition that has been worked out in centuries of cultural shift and apply it like a cookbook to a new discovery. You must be certain of the issues at stake before you go about condemning them." According to the views of many theologians and ethicists who have qualms about the cloning of humans, Wilmut's experiment could lead to an overturning of conventional ideas of motherhood, fatherhood and human identity; moreover, it raises new questions about the links between genetics and consciousness.

While it's unlikely that humans will anytime soon give birth to multiple genetic replicates of themselves, the cloning of the sheep named Dolly has added more urgency to a whole array of biomedical and moral issues, argued Lutheran theologian Ted Peters. "I'm not in favor of wildcat cloning, nor do I think it should be banned forever and ever," Peters said. "But if there's sufficient reason for caution, then we can wait and hear why we should go ahead."

Banning human cloning would not prevent the nightmare scenarios of totalitarian governments harvesting squadrons of cloned commandos genetically equipped with fighting instincts. But the real nightmare scenarios, in Peters's view, involve intentional or unintentional genetic damage done by private reproductive technology clinics that would sell cloning services. "That's where the energy will be put," he predicted. "And the outcome of that is an intuitive puzzle. It's possible that we would discover that cloning would not produce permanent psychological or physical damage to children. They could be like twins or triplets, genetically identical with their own unique consciousness and identity. But we don't really know what people who are in it for profit are going to be selling."

In Russell's view, the issues around cloning are more ethical than theological: Will the poor as well as the rich benefit from the fruits of research, be it milk from high-yield cows or access to replacement body parts? Will humans and animals risk losing their dignity and be reduced to commodities? Will the outcome benefit the common good, or fulfill the desires of an affluent few seeking immortality or the replication of a loved one?

Several ethicists agreed that human cloning is secondary to a more pressing genetic issue: the implications of human germ-line intervention—the alteration of defective genes that cause, for example, the hereditary disease cystic fibrosis. Scientists are close to perfecting techniques that would allow them to cure the disease not only in a human embryo but as a child matures to adulthood, in his or her offspring, generation after generation. The technique has the potential for great good—and massive harm. If a mistake were made in the manipulation of the gene, it would be impossible to correct in future generations. Germ-line intervention has been outlawed in Germany and several other European countries, but the U.S. has yet to develop policies for such procedures.

Rabbi Tendler Would Allow Some Cloning

Rabbi Moshe Tendler, a professor of microbiology who holds a chair in Jewish medical ethics at Yeshiva University in New York, has a scientist's appreciation of how cloning and germ-line intervention cut both ways on the ethical scale. "The biblical attitude is that God says 'Be fruitful and multiply, fill the Earth and master it. But the mastery is not over man: man is the forbidden fruit," Tendler said. "Only God can master man." To permanently modify a genotype has the capacity for evil, he said, because of the great risks involved. "Any error you make is permanent, eternal," he added. Similarly, implicit in the impulse to provide the childless with cloned offspring or to extend human longevity with cloned body parts is the evil of eugenics. "Declaring one human worthy of being replicated is to also declare other humans to be less worthy."

Though Tendler favors a ban on cloning for routine use as a method of reproduction, he suggests that there might be case-by-case exemptions from the ban. But he said he has little confidence that members of the scientific and medical communities have a highly enough evolved sense of ethics to use the new genetic technologies wisely. Remarked the rabbi: "Medical students listen to an ethics lecture from me for only one hour in six months of other classes. That's not enough."

Philip Boyle, senior vice-president of the Park Ridge Center for the Study of Health, Faith, and Ethics in Chicago expressed the fear that in the rush to cash in on this latest scientific advance, the Judeo-Christian tradition will be left in the dust. "These questions of the changing nature of parenthood and human identity have been around since the birth of the first test-tube baby in 1978," he noted. "But religions tend to cut a narrow swath. ... No one played out what it means to bring human life into existence, to take bone marrow and produce an image of ourselves." By remaining largely silent on these issues, Boyle said, organized religion is increasingly becoming a mere token voice in the ongoing public debate on emerging genetic technologies.

Meanwhile, President Clinton compared the recent scientific breakthroughs to the advancements that followed splitting the atom, unleashing the age of nuclear weapons. He urged scientists to "move with caution and care." In the wake of Wilmut's experiment Clinton asked the National Bioethics Advisory Commission to review what ramifications cloning would have for humans and report back to him in 90 days. The president also urged privately funded scientists not covered by the ban to observe a moratorium on human cloning experiments. Clinton said that while he understood the desire to use cloning as a means of curing disease, the procedure's ethical implications must be examined.

Two Republicans in Congress—Senator Christopher S. Bond of Missouri and Representative Vernon J. Ehlers of Michigan—have introduced legislation that would forbid all research on the cloning of human beings. But in testimony before Congress on March 12, several scientists emphasized the promise that such research holds in terms of curing disease and saving lives, and they urged Congress not to rush to ban it. Dolly cloner Wilmut, who opposes human cloning, testified that there is no immediate crisis because present methods of cloning are too inefficient to produce humans. Senator Tom Harkin (Democrat–Iowa) was alone among the legislators in predicting that human cloning would take place in his lifetime and declaring that he would welcome it.

31

Cloning Humans Should Never Be Attempted

John Frederic Kilner

Cigar, the champion racehorse, is a dud as a stud. Attempts to impregnate numerous mares have failed. But his handlers are not discouraged. They think they might try to have Cigar cloned.

If a sheep and a monkey can be cloned—and possibly a racehorse—can human clones be far behind? The process is novel, though the concept is not.

We have long known that virtually every cell of the body contains a person's complete genetic code. The exception is sperm or egg cells, each of which contains half the genetic material until the sperm fertilizes the egg and a new human being with a complete genetic code begins growing.

We have now learned that the partial genetic material in an unfertilized egg cell may be replaced by the complete genetic material from a cell taken from an adult. With a full genetic code, the egg cell behaves as if it has been fertilized. At least, that is how Dolly, the sheep cloned in Scotland, came to be. Hence, producing genetic copies of human beings now seems more likely.

We have been anticipating this possibility in humans for decades and have been playing with it in our imaginations. The movie *The Boys from Brazil* was about an attempt to clone Adolf Hitler. And in Aldous Huxley's novel *Brave New World*, clones were produced to fulfill undesirable social roles. More recently the movie *Multiplicity* portrayed a harried man who jumped at the chance to have himself copied—the better to tend to his office work, his home chores, and his family relationships. It all seems so attractive, at first glance, in our hectic, achievement-crazed society.

This appeared as "Stop Cloning Around" in *Christianity Today*, v41 (Apr. 28 1997). pp.10–11.

The Cost of Cloning Humans is Too High

But how do we achieve this technologically blissful state? *Multiplicity* is silent on this matter, implying that technique is best left to scientists, as if the rest of us are interested only in the outcome. But the experiments of Nazi Germany and the resulting Nuremberg Trials and Code taught us long ago that there is some knowledge that we must not pursue if it requires the use of immoral means.

The research necessary to develop human cloning will cause the deaths of human beings. Such deaths make the cost unacceptably high. In the process used to clone sheep, there were 277 failed attempts—including the deaths of several defective clones. In the monkey-cloning process, a living embryo was intentionally destroyed by taking the genetic material from the embryo's eight cells and inserting it into eight egg cells whose partial genetic material had been removed. Human embryos and human infants would likewise be lost as the technique is adapted to our own race.

Cloning May Be Acceptable for Crops and Livestock, but Is Wrong for Humans

Yet, as we press toward this new mark, we must ask: Is the production of human clones even a worthwhile goal? As movies and novels suggest, and godly wisdom confirms, human cloning is something neither to fool around with nor to attempt.

Cloning typically involves genetically copying some living thing for a particular purpose—a wheat plant that yields much grain, a cow that provides excellent milk. Such utilitarian approaches may be fine for cows and corn, but human beings, made in the image of God, have a God-given dignity that prevents us from regarding other people merely as means to fulfill our desires. We must not, for instance, produce clones with low intelligence (or low ambition) to provide menial labor, or produce clones to provide transplantable organs (their identical genetic code would minimize organ rejection). We should not even clone a child who dies tragically in order to remove the parents' grief, as if the clone could actually be the child who died.

All people are special creations of God who should be loved and respected as such. We must not demean them by fundamentally subordinating their interests to those of others.

There is a host of problems with human cloning that we

have yet to address. Who are the parents of a clone produced in a laboratory? The donor of the genetic material? The donor of the egg into which the material is transferred? The scientist who manipulates cells from anonymous donors? Who will provide the necessary love and care for this embryo, fetus, and then child—especially when mistakes are made and it would be easier simply to discard it?

The problems become legion when having children is removed from the context of marriage and even from responsible parenthood. For instance, Hope College's Allen Verhey asks "whether parenting is properly considered making children to match a specific design, as is clearly the case with cloning, or whether parenting is properly regarded as a disposition to be hospitable to children as given." Clearly, from a biblical perspective, it is the latter.

Further, the Bible portrays children as the fruit of a one-flesh love relationship, and for good reason. It is a context in which children flourish—in which their full humanity, material and nonmaterial, is respected and nourished. Those who provide them with physical (genetic) life also care for their ongoing physical as well as nonphysical needs.

As Valparaiso University's Gilbert Meilaender told CT, this further separation of procreating from marriage is bad for children. "The child inevitably becomes a product," says ethicist Meilaender, someone who is made, not begotten.

"To beget a child is to give birth to one who is like us, equal in dignity, for whom we care, but whose being we do not simply control. To 'make' a child is to create a product whose destiny we may well think we can shape. Hence, the 'begotten', not made language of the creed is relevant also to our understanding of the child and of the relation between the generations.

"If our purpose is to clone people as possible sources of perfectly matching organs," says Meilaender, "that clearly shows how we could come to regard the clone as a being we control—as simply an 'ensemble' of parts or organs."

Technology Should Not Conflict with God's Purpose

It is all too easy to lose sight of the fact that people are more than just physical beings, Meilaender's ensembles of organs. What most excites many people about cloning is the possibility of duplicate Michael Jordans, Mother Teresas, or Colin Powells.

However, were clones of any of these heroes to begin growing today, those clones would not turn out to be our heroes, for our heroes are not who they are simply because of their DNA. They, like us, were shaped by genetics and environment alike, with the spiritual capacity to evaluate, disregard, and at times to overcome either or both. Each clone would be subject to a unique set of environmental influences, and our loving God would surely accord each a unique personal relationship with him.

The problem with cloning is not the mere fact that technology is involved. Technology can help us do better what God has for us to do. The problem arises when we use technology for purposes that conflict with God's. And, as C. S. Lewis argued, technology never merely represents human mastery over nature; it also involves the power of some people over other people. This is as true in the genetic revolution as it was in the Industrial Revolution. When human cloning becomes technically possible, who will control who clones whom and for what ends? Like nuclear weaponry, the power to clone in the "wrong hands" could have devastating consequences.

There is wisdom in President Clinton's immediate move to forestall human cloning research until public debate and expert testimony have been digested and policies formulated. But there is even greater wisdom in never setting foot on the path that leads from brave new sheep to made-to-order organ donors, industrial drones, and vanity children.

32

Cloning Humans is Immoral

Jean Bethke Elshtain

As you've probably heard by now, Chicago physicist Richard Seed wants to clone a human being sometime over the next year and a half. He does not merely want it for the good of science, mind you, he wants it for his own glory as well. He wants to be the first. He wants to win a Nobel Prize. "The first person to produce a healthy human clone will be the winner of the Super Bowl," he gushes. "The second person will be the loser." Nothing will stand in Seed's way—not even the law. If the U.S. won't let him conduct his experiment, then, he insists, he will go offshore and clone people in Mexico.

These are audacious statements, to be sure. But they are not as audacious as Seed's other claim: namely, that God supports his endeavor because God wants us to be just like him, and God gave us the power to clone. One wonders where Seed learned his theology. In the Hebrew and Christian traditions, Seed's presumptive reference point, we are called to be cocreators, to participate in creation rather than to dominate and attain mastery over it. The Judeo-Christian tradition is all about acting morally with a humble sense of limit. To seek to be identical to God is idolatry; it is, quite literally, the sin of pride.

Seed has also speculated about introducing immortality into human life, so we could all live forever, "like God." But God doesn't "live" forever. God has no beginning or end. And God is not just a magnified form of a human being. Seed is guilty of what Vaclav Havel calls the "arrogant anthropocentrism" that has so ravaged this century.

Seed is blithely unaware of this—or, at least, he doesn't acknowledge it. And, even scarier, people seem to be playing along. Having defined God in human terms (God is like us, only more so), many Americans are convinced that we should propel

---

This appeared as "Bad Seed" in *The New Republic*, v218 n6, p.9, Feb 9, 1998. New Republic, Inc. 1998.

ourselves forward with nary a backward glance. If we can do it, we must do it. Meanwhile, a corollary to this belief has sprung up: If we can do it, why not do it? This, in sum, is the argument of Harvard Law Professor Laurence Tribe, who opined in The *New York Times* that he hadn't heard a convincing argument against cloning, and besides, it's a human right.

Maybe there's something appealing about such "freedom." But where does it take us? We've become so entranced by the notion that the self is merely the sum total of a person's choices that we have lost any moral framework for evaluating those choices. The upshot is a form of self-assertion that lays claim to human mastery and ambition—and, in so doing, creates a spiritual wasteland.

Indeed, such an understanding of persons lies at the foundation of two ideological projects that bedevil modern society. Those projects are untrammeled individualism, on one hand, and collectivism, on the other. Both projects see order as something humans must impose themselves. In the latter instance the imposition of order comes from above, once the "iron band of totalitarianism" (to use Hannah Arendt's memorable phrase) has forced people together. Its moral flaws are self-evident.

The rule of sovereign selves, however, is harder to criticize. This is because we inhabit a world in which we expect the concatenation of individual human choices to lead, automatically, to a moral end. We want to do it; therefore, we should do it.

But this is as ludicrous as the notion that unregulated markets will maximize positive good if left to their own devices. There is a powerful theoretical and social monism at work, fueling a project that celebrates human freedom but erodes our capacity to evaluate its consequences, thus destroying authentic moral freedom and responsibility. In a world in which the market reigns in all things, everything, finally, has the same value. We lose the ability to distinguish between one act or another; one claim or another; one argument or another.

Seed may not be particularly concerned with ethical and moral restraint anyway, but he does deny, vehemently, that his plan would be unethical or immoral. He would have the world believe his is an act of benevolence toward the unfortunate. Seed claims that infertile couples the world over are beating down his door, swamping his e-mail, clamoring: "Don't let them stop you." As he describes it, those who would stand in his way must be churlish, awful persons whose hearts harden when confronted with desperation.

But is it better to turn children into objects and products? Better to risk the threat of a damaging biogenetic uniformity, since much of the basic genetic information that goes into the creation of a child from two parents emerges as a result of sexual reproduction, something not replicable by definition when you pick one parent to clone? Better to embark on an experimental course that would likely result in unused "products" that didn't quite make it to the end of the conveyor belt and that no one wants?

No, there should be no Nobel Prize for this. Perhaps Seed can collaborate with Dr. Kevorkian and create a Mexico–Michigan axis. The creatures nobody wants can be shipped to Kevorkian for easy dispatch to spare their suffering. That will make more room for the Seed clones. Seed and Kevorkian are the Bobbsey Twins of our moral disorder.

33

The Judeo-Christian Ethic Opposes Cloning

Stephen Garrard Post

For purposes of discussion, I will assume that the cloning of humans is technologically possible. This supposition raises Einstein's concern: "Perfection of means and confusion of ends seems to characterize our age." Public reaction to human cloning has been strongly negative, although without much clear articulation as to why. My task is the Socratic one of helping to make explicit what is implicit in this uneasiness.

Some extremely hypothetical scenarios might be raised as if to justify human cloning. One might speculate, for example: If environmental toxins or pathogens should result in massive human infertility, human cloning might be imperative for species survival. But in fact recent claims about increasing male infertility worldwide have been found to be false. Some apologists for human cloning will insist on other strained "What if's." "What if" parents want to replace a dead child with an image of that child? "What if" we can enhance the human condition by cloning the "best" among us?

I shall offer seven unhypothetical criticisms of human cloning, but in no particular priority. The final criticism, however, is the chief one to which all else serves as preamble.

1. The Newness of Life. Although human cloning, if possible, is surely a novelty, it does not corner the market on newness. For millennia mothers and fathers have marveled at the newness of form in their newborns. I have watched newness unfold in our own two children, wonderful blends of the Amerasian variety. True, there probably is, as Freud argued, a certain narcissism in parental love, for we do see our own form partly reflected in the child, but, importantly, never entirely so.

This appeared as "The Judeo-Christian Case Against Human Cloning" in *America*, v176 (June 21–28 1997). pp. 19-22.

Sameness is dull, and as the French say, Vive la difference. It is possible that underlying the mystery of this newness of form is a creative wisdom that we humans will never quite equal.

This concern with the newness of each human form (identical twins are new genetic combinations as well) is not itself new. The scholar of constitutional law Laurence Tribe pointed out in 1978, for example, that human cloning could "alter the very meaning of humanity." Specifically, the cloned person would be "denied a sense of uniqueness." Let us remember that there is no strong analogy between human cloning and natural identical twinning, for in the latter case there is still the blessing of newness in the newborns, though they be two or more. While identical twins do occur naturally and are unique persons, this does not justify the temptation to impose external sameness more widely.

Sidney Callahan, a thoughtful psychologist, argues that the random fusion of a couple's genetic heritage "gives enough distance to allow the child also to be seen as a separate other," and she adds that the egoistic intent to deny uniqueness is wrong because ultimately depriving. By having a different form from that of either parent, I am visually a separate creature, and this contributes to the moral purpose of not reducing me to a mere copy utterly controlled by the purposes of a mother or father.

Perhaps human clones will not look exactly alike anyway, given the complex factors influencing genetic imprinting, as well as environmental factors affecting gene expression. But they will look more or less the same, rather than more or less different.

Surely no scientist would doubt that genetic diversity produced by procreation between a man and a woman will always be preferable to cloning, because procreation reduces the possibility for species annihilation through particular diseases or pathogens. Even in the absence of such pathogens, cloning means the loss of what geneticists describe as the additional hybrid vigor of new genetic combinations.

2. Making Males Reproductively Obsolete. Cloning requires human eggs, nuclei and uteri, all of which can be supplied by women. This makes males reproductively obsolete. This does not quite measure up to Shulamith Firestone's notion that women will only be able to free themselves from patriarchy through the eventual development of the artificial womb, but of course, with no men available, patriarchy ends–period.

Cloning, in the words of Richard McCormick, S.J., "would

involve removing insemination and fertilization from the marriage relationship, and it would also remove one of the partners from the entire process." Well, removal of social fatherhood is already a fait accompli in a culture of illegitimacy chic, and one to which some fertility clinics already marvelously contribute through artificial insemination by donor for single women. Removing male impregnators from the procreative dyad would simply drive the nail into the coffin of fatherhood, unless one thinks that biological and social fatherhood are utterly disconnected. Social fatherhood would still be possible in a world of clones, but this will lack the feature of participation in a continued biological lineage that seems to strengthen social fatherhood in general.

3. Under My Thumb: Cookie Cutters and Power. It is impossible to separate human cloning from concerns about power. There is the power of one generation over the external form of another, imposing the vicissitudes of one generation's fleeting image of the good upon the nature and destiny of the next. One need only peruse the innumerable texts on eugenics written by American geneticists in the 1920's to understand the arrogance of such visions.

One generation always influences the next in various ways, of course. But when one generation can, by the power of genetics, in the words of C. S. Lewis, "make its descendants what it pleases, all men who live after it are the patients of that power." What might our medicalized culture's images of human perfection become? In Lewis' words again, "For the power of Man to make himself what he pleases means, as we have seen, the power of some men to make other men what they please."

A certain amount of negative eugenics by prenatal testing and selective abortion is already established in American obstetrics. Cloning extends this power from the negative to the positive, and it is therefore even more foreboding.

This concern with overcontrol and overpower may be overstated because the relationship between genotype and realized social role remains highly obscure. Social role seems to be arrived at as much through luck and perseverance as anything else, although some innate capacities exist as genetically informed baselines.

4. Born to Be Harvested. One hears regularly that human clones would make good organ donors. But we ought not to presume that anyone wishes to give away body parts. The

assumption that the clone would choose to give body parts is completely unfounded. Forcing such a harvest would reduce the clone to a mere object for the use of others. A human person is an individual substance of a rational nature not to be treated as object, even if for one's own narcissistic gratification, let alone to procure organs. I have never been convinced that there are any ethical duties to donate organs.

5. The Problem of Mishaps. Dolly the celebrated ewe represents one success out of 277 embryos, nine of which were implanted. Only Dolly survived. While I do not wish to address here the issue of the moral status of the entity within the womb, suffice it to note that in this country there are many who would consider proposed research to clone humans as far too risky with regard to induced genetic defects. Embryo research in general is a matter of serious moral debate in the United States, and cloning will simply bring this to a head.

As one recent British expert on fertility studies writes, "Many of the animal clones that have been produced show serious developmental abnormalities, and, apart from ethical considerations, doctors would not run the medico-legal risks involved."

6. Sources of the Self. Presumably no one needs to be reminded that the self is formed by experience, environment and nurture. From a moral perspective, images of human goodness are largely virtue-based and therefore characterological. Aristotle and Thomas Aquinas believed that a good life is one in which, at one's last breath, one has a sense of integrity and meaning. Classically the shaping of human fulfillment has generally been a matter of negotiating with frailty and suffering through perseverance in order to build character. It is not the earthen vessels, but the treasure within them that counts. A self is not so much a genotype as a life journey. Martin Luther King Jr. was getting at this when he said that the content of character is more important than the color of skin.

The very idea of cloning tends to focus images of the good self on the physiological substrate, not on the journey of life and our responses to it, some of them compensations to purported "imperfections" in the vessel. The idea of the designer baby will emerge, as though external form is as important as the inner self.

7. Respect for Nature and Nature's God. In the words of Jewish bioethicist Fred Rosner, cloning goes so far in violating the

structure of nature that it can be considered as "encroaching on the Creator's domain." Is the union of sex, marriage, love and procreation something to dismiss lightly?

Marriage is the union of female and male that alone allows for procreation in which children can benefit developmentally from both a mother and father. In the Gospel of Mark, Jesus draws on ancient Jewish teachings when he asserts, "Therefore what God has joined together, let no man separate." Regardless of the degree of extendedness in any family, there remains the core nucleus: wife, husband and children. Yet the nucleus can be split by various cultural forces (e.g., infidelity as interesting, illegitimacy as chic), poverty, patriarchal violence and now cloning.

A cursory study of the Hebrew Bible shows the exuberant and immensely powerful statements of Genesis 1, in which a purposeful, ordering God pronounces that all stages of creation are "good." The text proclaims, "So God created humankind in his image, in the image of God he created them, male and female he created them" (Gen. 1: 27). This God commands the couple, each equally in God's likeness, to "be fruitful and multiply." The divine prototype was thus established at the very outset of the Hebrew Bible: "Therefore a man leaves his father and his mother and clings to his wife, and they become one flesh" (Gen. 2: 24).

The dominant theme of Genesis 1 is creative intention. God creates, and what is created procreates, thereby ensuring the continued presence of God's creation. The creation of man and woman is good in part because it will endure. Catholic natural law ethicists and Protestant proponents of "orders of creation" alike find divine will and principle in the passages of Genesis 1.

A major study on the family by the Christian ethicist Max Stackhouse suggests that just as the pre-Socratic philosophers discovered still valid truths about geometry, so the biblical authors of Chapters One and Two of Genesis "saw something of the basic design, purpose, and context of life that transcends every sociohistorical epoch." Specifically, this design includes "fidelity in communion" between male and female oriented toward "generativity" and an enduring family the precise social details of which are worked out in the context of political economies.

Christianity appropriated the Hebrew Bible and had its origin in a Jew from Nazareth and his Jewish followers. The basic contours of Christian thought on marriage and family therefore owe a great deal to Judaism. These Hebraic roots that shape the words of Jesus stand within Malachi's prophetic tradition of emphasis on inviolable monogamy. In Mk. 10: 2-12 we read:

> The Pharisees approached and asked, "Is it lawful for a husband to divorce his wife?" They were testing him. He said to them in reply, "What did Moses command you?" They replied, "Moses permitted him to write a bill of divorce and dismiss her." But Jesus told them, "Because of the hardness of your hearts he wrote you this commandment. But from the beginning of creation, 'God made them male and female. For this reason a man shall leave his father and mother (and be joined to his wife), and the two shall become one flesh. So they are no longer two but one flesh. Therefore what God has joined together, no human being must separate." In the house the disciples again questioned him about this. He said to them, "Whoever divorces his wife and marries another commits adultery against her; and if she divorces her husband and marries another, she commits adultery."

Here Jesus quotes Gen. 1: 27 ("God made them male and female") and Gen. 2: 24 ("the two shall become one flesh").

Christians side with the deep wisdom of the teachings of Jesus, manifest in a thoughtful respect for the laws of nature that reflect the word of God. Christians simply cannot and must not underestimate the threat of human cloning to unravel what is both naturally and eternally good.

34

Human Cloning Is Not Against the Will of God

Gregory E. Pence

The most general argument against creating an adult human being by NST [Nuclear Somatic Transfer] is that it is against the will of God. Some people believe that humans have no right to change the way that humans are created because sexual reproduction was ordained by God and that to try to change this way is sinful.

Upon hearing the news of Dolly, Duke University divinity professor Stanley Hauerwas said that those who wanted to clone Dolly "are going to try to sell it with wonderful benefits" to medical and animal industries. He condemned the procedure because he thought it was "a kind of drive behind this for us to be our own creators," raising the images of Drs. Frankenstein and human hubris. A scant ten days after Dolly's announcement, the Christian Life Commission of the Southern Baptist Convention predictably called for a federal law against human cloning, as well as an international law to the same effect.[1] The Vatican in April of 1997 asserted: "the right to be born in a human way." It urged all nations to ban human cloning and, in manner typical of Vatican ethics, urged them to make no exceptions.[2] An enlightened Shi'ite Muslim Jurist named Sheikh Fadlallah suggested that originating a child by cloning should be punished by death or, at the very least, amputation.[3]

The NBAC *Report* quoted Paul Ramsey's dictum that, "Religious people have never denied, indeed they affirm, that God means to kill us all in the end, and in the end he is going to succeed."[4] As said, testifying before NBAC, Lutheran theologian

---

This appeared in Gregory E. Pence, *Who's Afraid of Human Cloning?* (Lanham, Md: Rowman and Littlefield, 1998), pp. 119–122. Reprinted by permission of Rowman and Littlefield.

Gilbert Meilaender emphasized the importance of "the creation story in the first chapter of Genesis [that] depicts the creation of humankind as male and female, sexually differentiated and enjoined by God's grace to sustain human life through procreation."[5] Other theologians testifying emphasized the Scriptural basis of: God's creation of humans, warnings not to play God, dangers of the quest for knowledge, the need for responsible dominion over nature, and the ultimate folly of human destiny.[6]

Verses 26–27 of the first chapter of Genesis tell us, "And God said, Let us make man in our image, after our likeness ... So God created man in his own image, in the image of God created he them; male and female created he them." Two chapters later, we get another famous, relevant story:

> 6. And when the woman saw that the tree was good for food, and that it was pleasant to the eyes, and a tree to be desired to make one wise, she took of the fruit thereof, and did eat, and gave also unto her husband with her; and he did eat.
>
> 7. And the eyes of them both were opened, and they knew that they were naked; and they sewed fig leaves together, and made themselves aprons.

This story tells an oft-told tale in the Old Testament that once-upon-a-time there was a time of moral perfection, but because of some bad act of humans, things changed and disaster followed. Genesis describes Eden as a pure and simple land that ultimately gets so corrupted that God brings the great flood, allowing only Noah and his family to escape. Often the agent of corruption is the human quest for knowledge, represented by the Tower of Babel, which also signifies human hubris before God.

These stories are our most ancient tales and, as such, resonate deeply within us. They convey the message that a small bit of new knowledge can be treacherous for humanity. Humanity is forbidden from knowing certain things, and the consequence of knowing these things is that a black cloak of doom will cover our world.

Despite these deeply-ingrained feelings, it is a grave mistake to let such ancient stories dictate modern public policy. First, and most obviously, they are just stories. Many of the events described in Genesis, especially the story of the Garden of Eden, are not actual events but symbolic fables, repeated in many Near

Eastern cultures. For example, pre-Hebraic civilizations such as Babylonia also had the story of a great flood.

Second, and of equal importance, the Old Testament is not full of moral wisdom. Although Meilaender's entire testimony derives from these Scriptural stories, like all such references to Scripture, he only goes there to justify what he already believes. If a neutral person merely read the Old Testament and followed what it said, he would find that: Abraham and King David make women captured in war into slaves in their harems; Exodus 21:2 condones human slavery; Jephthah's daughter is required as a human sacrifice; sex with slave-maids is permitted to create male heirs when Sarah's and Rachel's husbands are barren; and Exodus 23:27 permits killing women and children of enemies in war.

Genesis 3:16 also says famously, "Unto the woman he said, I will greatly multiply thy sorrow and thy conception; in sorrow thou shalt bring forth children; and thy desire shall be to thy husband, and he shall rule over thee." Obviously, this event could not have really happened and must just be a story, because otherwise we are forced to see God as utterly unjust. Not only is Eve punished for her own act, so are billions of women afterwards in human history who must experience painful childbirth. Although Scripture condones these injustices, Meilaender undoubtedly does not think them right, and that is because he has some non-Scriptural standard of right on which to base his judgment. It is also worth noting that rabbis such as Moshe Tendler, who also testified before NBAC and whose religion has been studying Genesis longer than Christianity, do not think that the creation story of Genesis implies an absolute ban on human cloning.

Consider also that the story of the Tree of Knowledge is about moral knowledge: "then your eyes shall be opened, and ye shall be as gods, knowing good and evil" (3:5). What should we infer from this story. That humans should know nothing about morality? Perhaps humans made no moral judgments in the Garden of Eden, but how is that story relevant to a world where humans must constantly do so?

To accept Old Testament morality in general is to accept a fatalistic worldview. That is understandable because the Old Testament was edited to its final form over two thousand years ago when being fatalistic made sense. The Old Testament did not anticipate genetic therapy, artificial skin, and organ transplants.

In more modern theology, it takes a mighty lot of

intermediate steps of interpretation and reasoning to get from (1) A rational God exists and cares about us, to (2) It is immoral to originate a human by NST. Such steps would have to counter the premises that, (3) Because a rational God exists and cares about us, He allows us to make new discoveries in medicine and science, and that, (4) Because God is rational and cares about us, he directs us to create humans in ways that are rational and that express caring about human beings. Substantial argument would be needed to prove that originating a child by NST by parents with good motives was against (4). I do not think that can be done.

After all the dust settles about originating babies by NST, it may turn out that opposition to it stems *only* from religious people or from those whose thinking is colored by traditional religious beliefs. If so, isn't there a problem with recognizing such views in American law? The U.S. Constitution forbids the establishment of laws and policies that are solely motivated by religious beliefs. Should studies in non-human mammalian embryogenesis indicate that NST would be safe in humans, and should well-motivated parents be indentified who had good reasons to try it (being at risk for having babies with genetic disease), would not further opposition be solely based on religious grounds? And as such, would it not violate our Constitution to have it ground a federal ban that restricts others' reproductive freedom?

References to Reading 34

1. National Bioethics Advisory Commission (NBAC), *Cloning Human Beings: Report and Recommendations of the National Bioethics Advisory Commission*, Rockville, Md., June 1997, 56.

2. NBAC, *Cloning Human Beings*, 56.

3. NBAC, *Cloning Human Beings*, 59.

4. Quoted in the NBAC, *Cloning Human Beings*, 47, from Paul Ramsey, *Fabricated Man: The Ethics of Genetic Control* (New Haven, Conn.: Yale University Press, 1997), 136.

5. Gilbert Meilaender, "Begetting and Cloning," *First Things* 74 (June/July 1997), 41–43.

6. NBAC, *Cloning Human Beings*, 43–49.

VII

Regulation of Reproduction or Reproductive Liberty?

Introduction to Part VII

Since it is commonly believed that individuals have the right to procreate, within the limits of voluntary consent, as they wish, some argue that this principle should extend to cloning, while others maintain that cloning is an area where individual freedom of choice ought not to prevail.

A statement by Diane Aronson, Executive Director of RESOLVE (Reading 35), voices that organization's opposition to attempts to ban embryo research. Thomas Shannon, however, contends (Reading 36) that individual choice cannot be assumed to be the only morally relevant value.

In Reading 37, Mark Eibert marshals several constitutional arguments to suport the view that individuals possess the right to clone themselves. Lori Andrews (Reading 38) maintains that, in the U.S., government controls over reproduction, and cloning in particular, are insufficient.

35

Statement by RESOLVE

Diane D. Aronson

RESOLVE expresses its strong support for ... [legislation] specifying that research using somatic cell nuclear transfer technology should not be banned while recommending a moratorium on the cloning of a human being until further review.

RESOLVE ... [does not support a] ban embryo research. Embryo research has been instrumental in the development of procedures that allow many couples to overcome the difficulties they experience as they strive to build families. The emotional and physical consequences of this struggle can be overwhelming. *In vitro* fertilization is an amazing technology which would not have been possible without the knowledge gained through embryo research. This effective treatment has brought about the birth of thousands of much-wanted babies. Continued embryo research has the potential to further the understanding of the causes of infertility, including the tragedy of miscarriage, as well as provide information which can lead to new breakthroughs.

As a national organization which provides support, advocacy and education to those experiencing infertility, RESOLVE is contacted by thousands of people from all walks of life who are struggling with this disease. The stories about their struggles can be heart-wrenching. The success stories about the joy and overwhelming appreciation of the children that are brought into this world are enormously heart-warming. Avenues for further research to help couples must not be halted. RESOLVE joins with many other organizations across the country in expressing its opposition to any attempts to ban embryo research. We applaud your efforts to develop carefully-constructed legislation which will not impact the potential for medical advances that will help the many couples struggling to build much-wanted families.

This appeared in the *Congressional Record* of February 10, 1998 p. S569.

36

When It Comes to Cloning, Personal Choice Should Not Be the Final Word

Thomas A. Shannon

Since the birth of Dolly, the cloned sheep, the debate over human cloning has been characterized by misunderstanding and exaggeration. These have been fueled in part by a lack of careful distinctions in popular reports and discourse. For example, three distinct types of cloning—gene cloning, cellular cloning, and whole-organism cloning—have sometimes been fused in media coverage, leading to widespread confusion. Gene cloning multiplies identical copies of various genes; cellular cloning, a more complicated technique, replicates whole cells; and whole-organism cloning—the most complicated—reproduces whole organisms.

Gene and cell cloning are well-established, standard biotechnical research methods and must be distinguished and discussed separately from organism cloning. Organism cloning, *à la* Dolly, signaled a dramatic scientific breakthrough because Dolly's cloning was accomplished with cells that were six years old and fully differentiated. The common wisdom until then was that such cells could not be reprogrammed to generate a new being. While Dolly has vast ramifications for both animal husbandry and the production of pharmaceuticals, the most important scientific breakthrough was the reactivation of fully differentiated cells, a point frequently missed by popular reports. Recently, this very breakthrough has been called into question by some senior scientists because the procedure has not yet been replicated, and because, in the absence of the ewe from which Dolly's cells were taken, there is no way of verifying that the cells that were used were adult cells rather than fetal cells (the same ewe was pregnant

---

This appeared as "Cloning Myths" in *Commonweal*, v125, 1998.

at the time).

Human cloning is an altogether different matter from the cloning of sheep and cattle. This fact has been pointed out not only by ethicists and government officials, but by research scientists. But as in the debate about the cloning of animals, the human cloning debate has been characterized by many misunderstandings. The most common has been that if I were to clone myself, I would create another me, because each of us would have the same genetic identity. On reflection, most would recognize that this kind of simplistic genetic reductionism can't be right: Identical twins—who share the same genetic identity—are not identical persons. Furthermore, such reductionism fails to take into account the importance of environment, experience, physical and psychological factors, and a host of other variables that contribute to an individual's formation and identity.

Another mistaken assumption is that human clones would necessarily act alike: All Michael Jordan clones would be superior basketball stars because of genetic identity. Again, this ignores not only the role of environment, but the constitutive part played by human freedom and endeavor in shaping our identity.

Finally, some have assumed that human cloning would lead to a ready supply of replacement parts to be made available for use by the cloner when, for example, one of his or her organs might fail. But such an assumption fails to acknowledge that a clone would also be a human person with inalienable rights and vital interests—not to mention moral and legal standing—in how his or her body ought to be used. Once we grant that, however, the replacement debate necessarily ends, but not the human cloning debate which now shifts to assisted reproduction.

Can (or should) a couple who have been unable to conceive in any other way that would guarantee a genetic relation to one of the parents be prohibited from using cloning to produce a child? Unless one has moral objections to all forms of assisted reproduction, the argument goes, cloning can be yet another morally valid technique for producing a child. While there is a certain logic to this train of thought, there are nonetheless three reasons why its conclusion should be resisted.

First, the argument is predicated on the sovereignty of personal choice: Someone wants the procedure, consents to it, can pay for it, and, therefore, ought to have it. Such an argument begs the question of the sufficiency of individual choice as the only morally relevant value in ethical analysis.

A second argument is a variant of personal choice which states that having a child is a strictly private matter and, therefore, should not be subjected to outside social analysis. But even though the choice is private and assisted-reproduction clinics are privately owned and funded, nonetheless such private choices in private institutions have profound social implications. For example, insurance and health-care interests already play a role in how such choices are mediated because some insurance plans cover some infertility services and health care during pregnancy, and thus increase the cost of premiums for all members of the plan. Furthermore, since multiple pregnancies are a possibility because multiple embryos are implanted to insure a pregnancy, there might be significant intensive-care-unit expenses incurred, a cost which will again be passed along to members of the plan. Such social costs render private choices less private and more open to social evaluation.

A third argument is that as in so many other aspects of American life, the market should rule. Eggs and sperm are currently available for sale (though the price of eggs has just more than doubled to $5,000), as are custom-designed embryos. The technology of cloning can further refine that product helping to market it better. But once again a deeper question has been avoided. For if we market ourselves in such a manner—even to fulfill a noble wish, such as the desire to have a child—and place a quantifiable price on the transaction, we have made ourselves into objects whose price is known but whose value has been forgotten. In so doing, we risk diminishing the value of human life.

The critical question before us, therefore, is whether we can apply the developments in cloning to reproductive technology in such a way as to assure the larger good of human dignity and individuality in a moral manner. To do that we will need both time and discernment. Insuring that we take the time and expend the effort to think hard about these scientific developments constitutes the most critical phase of the ongoing cloning debate.

37

Freedom to Reproduce is a Right

Mark D. Eibert

"There are some avenues that should be off limits to science. If scientists will not draw the line for themselves, it is up to the elected representatives of the people to draw it for them."

Thus declared Senator Christopher "Kit" Bond (Republican–Missouri) one of the sponsors of S. 1601, the official Republican bill to outlaw human cloning. The bill would impose a 10-year prison sentence on anyone who uses "human somatic cell nuclear transfer technology" to produce an embryo, even if only to study cloning in the laboratory. If enacted into law, the bill would effectively ban all research into the potential benefits of human cloning. Scientists who use the technology for any reason—and infertile women who use it to have children—would go to jail.

Not to be outdone, Democrats have come up with a competing bill. Senators Ted Kennedy (Massachusetts) and Dianne Feinstein (California) have proposed S. 1602, which would ban human cloning for at least 10 years. It would allow scientists to conduct limited experiments with cloning in the laboratory, provided any human embryos are destroyed at an early stage rather than implanted into a woman's uterus and allowed to be born.

If the experiment goes too far, the Kennedy–Feinstein bill would impose a $1 million fine and government confiscation of all property, real or personal, used in or derived from the experiment. The same penalties that apply to scientists appear to apply to new parents who might use the technology to have babies.

The near unanimity on Capitol Hill about the need to ban human cloning makes it likely that some sort of bill will be voted on this session and that it will seriously restrict scientists' ability to study human cloning. In the meantime, federal bureaucrats have leapt into the breach. In January, the U.S. Food and Drug

---

This appeared as "Clone Wars" in *Reason*. v30, 1998.

Administration announced that it planned to "regulate" (that is, prohibit) human cloning. In the past, the FDA has largely ignored the fertility industry, making no effort to regulate *in vitro* fertilization, methods for injecting sperm into eggs, and other advanced reproductive technologies that have much in common with cloning techniques.

An FDA spokesperson told me that although Congress never expressly granted the agency jurisdiction over cloning, the FDA can regulate it under its statutory authority over biological products (like vaccines or blood used in transfusions) and drugs. But even Representative Vernon Ehlers (Republican–Michigan), one of the most outspoken congressional opponents of cloning, admits that "it's hard to argue that a cloning procedure is a drug." Of course, even if Congress had granted the FDA explicit authority to regulate cloning, such authority would only be valid if Congress had the constitutional power to regulate reproduction—which is itself a highly questionable assumption (more on that later).

Nor have state legislatures been standing still. Effective January 1, California became the first state to outlaw human cloning. California's law defines "cloning" so broadly and inaccurately—as creating children by the transfer of nuclei from any type of cell to enucleated eggs—that it also bans a promising new infertility treatment that has nothing to do with cloning. In that new procedure, doctors transfer nuclei from older, dysfunctional eggs (not differentiated adult cells as in cloning) to young, healthy donor eggs, and then inseminate the eggs with the husband's sperm—thus conceiving an ordinary child bearing the genes of both parents. Taking California as their bellwether, many other states are poised to follow in passing very restrictive measures.

Constitutional Principles Conflict with Any Cloning Ban

What started this unprecedented governmental grab for power over both human reproduction and scientific inquiry? Within days after Dolly, the cloned sheep, made her debut, President Clinton publicly condemned human cloning. He opined that "any discovery that touches upon human creation is not simply a matter of scientific inquiry. It is a matter of morality and spirituality as well. Each human life is unique, born of a miracle that reaches beyond laboratory science."

Clinton then ordered his National Bioethics Advisory Commission to spend all of 90 days studying the issue—after which the board announced that it agreed with Clinton. Thus, Clinton succeeded in framing the debate this way: Human cloning was inherently bad, and the federal government had the power to outlaw it.

But in fact, it's far from clear that the government has such far-reaching authority. Several fundamental constitutional principles conflict with any cloning ban. Chief among them are the right of adults to have children and the right of scientists to investigate nature.

The Supreme Court has ruled that every American has a constitutional right to "bear or beget" children. This includes the right of infertile people to use sophisticated medical technologies like *in vitro* fertilization. As the U.S. District Court for the Northern District of Illinois explained, "Within the cluster of constitutionally protected choices that includes the right to have access to contraceptives, there must be included … the right to submit to a medical procedure that may bring about, rather than prevent, pregnancy."

About 15 percent of Americans are infertile, and doctors often cannot help them. Federal statistics show that *in vitro* fertilization and related technologies have an average national success rate of less than 20 percent. Similarly, a Consumer Reports study concludes that fertility clinics produce babies for only 25 percent of patients. That leaves millions of people who still cannot have children, often because they can't produce viable eggs or sperm, even with fertility drugs. Until recently, their only options have been to adopt or to use eggs or sperm donated by strangers.

Once cloning technology is perfected, however, infertile individuals will no longer need viable eggs or sperm to conceive their own genetic children—any body cell will do. Thus, cloning may soon offer many Americans the only way possible to exercise their constitutional right to reproduce. For them, cloning bans are the practical equivalent of forced sterilization.

In 1942, the Supreme Court struck down a law requiring the sterilization of convicted criminals, holding that procreation is "one of the basic civil rights of man," and that denying convicts the right to have children constitutes "irreparable injury" and "forever deprived them of a basic liberty." To uphold a cloning ban, then, a court would have to rule that naturally infertile citizens have less right to try to have children than convicted rapists and child molesters do.

Many politicians and bureaucrats who want to ban human cloning say they need their new powers to "protect" children. Reciting a long list of speculative harms, ranging from possible physical deformities to the psychic pain of being an identical twin, they argue in essence that cloned children would be better off never being born at all.

But politicians and the media have grossly overstated the physical dangers of cloning. The Dolly experiment started with 277 fused eggs, of which only 29 became embryos. All the embryos were transferred to 13 sheep. One became pregnant, with Dolly. The success rate per uterine transfer (one perfect offspring from 13 sheep, with no miscarriages) was better than the early success rates for *in vitro* fertilization. Subsequent animal experiments in Wisconsin have already made the process much more efficient, and improvements will presumably continue as long as further research is allowed.

Government Control over Reproduction Goes Against American Tradition

More fundamentally, the government does not have the constitutional authority to decide who gets born—although it once thought it did, a period that constitutes a dark chapter of our national heritage. In the early twentieth century, 30 states adopted eugenics laws, which required citizens with conditions thought to be inheritable (insanity, criminal tendencies, retardation, epilepsy, etc.) to be sterilized—partly as a means of "protecting" the unfortunate children from being born.

In 1927, the U.S. Supreme Court upheld such a law, with Oliver Wendell Holmes writing for the majority, "It would be better for all the world, if instead of waiting to execute degenerate offspring for crime, or to let them starve for their imbecility, society can prevent those who are manifestly unfit from continuing their kind ... Three generations of imbeciles are enough."

California's eugenics law in particular was admired and emulated in other countries—including Germany in 1933. But during and after World War II, when Americans learned how the Nazis had used their power to decide who was "perfect" enough to be born, public and judicial opinion about eugenics began to shift. By the 1960s, most of the eugenics laws in this country had been either repealed, fallen into disuse, or were struck down as violating constitutional guarantees of due process and equal protection.

Indeed, those old eugenics laws were a brief deviation from an American tradition that has otherwise been unbroken for over 200 years. In America, it has always been the prospective parents, never the government, who decided how much risk was acceptable for a mother and her baby—even where the potential harm was much more certain and serious than anything threatened by cloning.

Hence, *in vitro* fertilization and fertility drugs are legal, even though they create much higher risks of miscarriages, multiple births, and associated birth defects. Individuals who themselves have or are known carriers of serious inheritable mental or physical defects such as sickle cell anemia, hemophilia, cystic fibrosis, muscular dystrophy, and Tay-Sachs disease are allowed to reproduce, naturally and through *in vitro* fertilization, even though they risk having babies with serious, or even fatal, defects or diseases. Older mothers at risk of having babies with Down Syndrome, and even women with AIDS, are also allowed to reproduce, both naturally and artificially. Even if prenatal testing shows a fetus to have a serious defect like Down Syndrome, no law requires the parents to abort it to save it from a life of suffering.

In short, until science revealed that human cloning was possible, society assumed that prospective parents could decide for themselves and their unborn children how much risk and suffering were an acceptable part of life. But in the brave new world of the federal bureaucrat, that assumption no longer holds true.

Ironically, some cloning opponents have turned the eugenics argument on its head, contending that cloning could lead to "designer children" and superior beings who might one day rule mankind. But allowing infertile individuals to conceive children whose genome is nearly identical to their already existing genomes no more creates "designer children" that it creates "designer parents." More important, these opponents miss the point that only government has the broad coercive power over society as a whole necessary to make eugenics laws aimed at "improving the race." It is those who support laws to ban cloning who are in effect urging the passage of a new eugenics law, not those who want to keep government out of the business of deciding who is perfect enough or socially desirable enough to be born.

Another significant driving force behind attempts to restrict or reverse an expansion in human knowledge stems from religious convictions. Interestingly, there is no necessary theological opposition to cloning: For example, two leading rabbis and a Muslim scholar who testified before the National Bioethics Advisory Commission had no objection to the practice and even advanced religious arguments for cloning.

Still, politicians from both major parties have already advanced religious arguments against cloning. President Clinton wants to outlaw cloning as a challenge to "our cherished concepts of faith and humanity." House Majority Leader Dick Armey also opposes cloning, saying that "to be human is to be made in the image and likeness of a loving God," and that "creating multiple copies of God's unique handiwork" is bad for a variety of reasons. Senator Bond warns that "humans are not God and they should not be allowed to play God"—a formulation similar to that of Albert Moraczewski, a theologian with the National Conference of Catholic Bishops, who told the president's commission, "Cloning exceeds the limits of the delegated dominions given to the human race."

Earlier Advances Were Attacked on Similar Grounds

Of course, virtually every major medical, scientific, and technological advance in modern history was initially criticized as "playing God." To give just two recent examples, heart transplants and "test tube babies" both faced religious opposition when first introduced. Today, heart transplants save 2,000 lives every year, and *in vitro* fertilization helped infertile Americans have 11,000 babies in 1995 alone.

Religious belief doesn't require opposition to these sorts of expansions in human knowledge and technology. And basing a cloning ban primarily on religious grounds would seem to violate the Establishment Clause. But that's not the only potential constitutional problem with a ban.

Many courts and commentators say that a constitutional right of scientific inquiry is inherent in the rights of free speech and personal liberty. To be sure, certain governmental attempts to restrict the methods scientists can use have been upheld—for example, regulations requiring free and informed consent by experimental subjects. But those have to do with protecting the rights of others. Cloning bans try to stop research that everyone directly concerned wants to continue. As one member of the National Bioethics Advisory Commission observed, if the group's recommendation to ban cloning is enacted, it would apparently be the first time in American history that an entire field of medical research has been outlawed.

Prohibiting scientific and medical activities would also raise troubling enforcement issues. How exactly would the FBI—in its new role as "reproductive police" and scientific overseer—learn, then prove, that scientists, physicians, or parents were violating the ban? Would they raid research laboratories and universities? Seize and read the private medical records of infertility patients? Burst into operating rooms with their guns drawn? Grill new mothers about how their babies were conceived? Offer doctors reduced sentences for testifying against the patients whose babies they delivered? And would the government really confiscate, say, Stanford University Medical Center, if one of its many researchers or clinicians "goes too far"?

The year since the announcement of Dolly's birth has seen unprecedented efforts by government to expand its power over both human reproduction and science. Decisions traditionally made by individuals—such as whether and how to have children, or to study the secrets of nature—have suddenly been recast as political decisions to be made in Washington.

The Nightmares Are Based on Misunderstandings

Human cloning, when it actually arrives, is not apt to have dire consequences. Children conceived through cloning technology will be not "Xerox copies" but unique individuals with their own personalities and full human rights. Once this basic fact is understood, the only people likely to be interested in creating children through cloning technology are incurably infertile individuals. There are already tens of millions of identical twins walking the earth, and they have posed no threat so far to God, the family, or country. A few more twins, born to parents who desperately want to have, raise, and love them as their own children, will hardly be noticed.

As for the nightmare fantasies spun by cloning opponents, even the president's special commission has admitted that fears of cloning being used to create hordes of Hitlers or armies of identical slaves are "based … on gross misunderstandings of human biology and psychology." And laws already prohibit criminal masterminds from holding slaves, abusing children, or cutting up people for spare body parts.

As harmless as the fact of cloning may be, the fear of cloning is already bearing bitter fruit: unprecedented extensions of government power, based either on unlikely nightmare scenarios or on an unreasoning fear that humans were "not meant" to know or do certain things. Far from protecting the "sanctity" of human life, such an attitude, if consistently applied, would doom the human race to a "natural" state of misery.

38

We Need Regulation of Reproduction

Lori B. Andrews

When the physicist Richard Seed announced in December that he intended to clone humans, he was denounced by President Clinton and labeled a "mad scientist" by Donna Shalala, Secretary of Health and Human Services. Yet his controversial plan, along with recent advances in animal cloning, indicates that we may be closer to the cloning of humans than we previously had thought. Seed's announcement, and the subsequent scramble by policy makers—from state legislators to the Acting Commissioner of the Food and Drug Administration—to stop him, points out major problems in our current approach to regulating technologies used in human reproduction.

The United States has at least 281 *in vitro* fertilization clinics, according to a survey released in December by the Centers for Disease Control. At least half of the clinics have the equipment and personnel necessary to undertake human cloning. Many of them offer a procedure known as "intracytoplasmic sperm injection," in which an embryologist uses a microscopic pipette to inject a single sperm into a woman's egg. Clinics use this approach to treat infertile couples when the man has a low sperm count, but the clinics could easily adapt it to clone humans, by injecting nucleic DNA from an adult's cell into an egg from which the nucleus has been removed.

Although President Clinton has issued an order forbidding the use of federal funds for human cloning, that ban will have little effect on fertility clinics. For 20 years, the federal government has refused to provide funds for research on *in vitro* fertilization, but that hasn't stopped the hundreds of privately financed I.V.F.

This appeared as ""Human Cloning: Assessing the Ethical and Legal Quandaries" in *Chronicle of Higher Education*, v44 n23, pp. B4–B5, February 13, 1998,

clinics from creating tens of thousands of babies, with an average success rate now of 20 percent. And the President's ban won't stop Richard Seed; the Raelians, a Swedish religious group, have offered Seed private funds and laboratory space to begin his work.

The policy makers bent on stopping Seed have failed to examine the larger issues already raised by the unregulated introduction of new reproductive technologies. And laws being proposed by legislators at the federal and state levels to deal with cloning are rapidly becoming obsolete, as scientific developments change the terms of the debate. It is time to consider exactly what troubles us about the idea of creating children through human cloning, determine if some of those same issues affect the operations of existing fertility clinics, and develop laws and regulations for both new and existing reproductive technologies.

In most states, fertility clinics are observing a voluntary moratorium on human cloning, because they suspect that success rates would be low, and because research on animals suggests the possibility of physical risks to any offspring in cloning attempts. After all, of the 277 attempts by the Scottish embryologist Ian Wilmut to create a clone from an adult sheep's mammary cell, only one produced a live offspring—Dolly. Most people believe that it is unethical to subject humans to a procedure with those odds; it would also be prohibitively expensive. Few women now make their eggs available to fertility clinics, and when they do, they are paid approximately $2,000 per egg. No clinic is going to be able to acquire 277 eggs for use with a single patient; even if one did, 277 eggs at $2,000 apiece would require $554,000 for each child created through cloning.

Human Cloning May Become Cheaper and Surer

Recent scientific developments are changing this equation, though. At a meeting of the International Embryo Transfer Society last month, two University of Massachusetts scientists, James Robl and Steven Stice, announced a new technique for producing clones of adult cows, with a success rate of 5 to 10 percent. They added cells that were dividing normally to cow eggs whose nuclei had been removed, and then they waited longer to reactivate the DNA than Ian Wilmut did in the experiments that produced Dolly. The success rate of this new cloning technique is comparable to the success rates of 13 percent when clinics use frozen human embryos in fertility treatments and 7 percent for *in vitro* fertilization attempts in women over age 39.

At the same meeting of the embryo-transfer society, the reproductive biologists Neal First, of the University of Wisconsin, and Tanja Dominko, of the Oregon Regional Primate Research Center, announced another development in cloning. They had inserted nucleic DNA from several species—rats, sheep, pigs, and rhesus monkeys—into cow eggs whose own nuclei had been removed, and the eggs activated the nucleic DNA to produce a clone of the donor of the DNA. Although all of the embryos that First and Dominko have created with this technique so far have miscarried, their work raises the theoretical possibility that cow eggs could be used as a universal incubator for any adult mammal's cell. Using cow eggs to clone humans means that women would not have to be subjected to the physical risks of donating their eggs. And cow eggs would make the procedure less expensive. Neal First got his eggs from a slaughterhouse.

If, as appears possible, experiments eventually indicate that cloning would be as effective as, and less expensive than, the current reproductive technologies, then infertility doctors may feel that it is ethical to try to clone humans. There still will be physical risks, as indicated by the cloned sheep, frogs, and cows that have been born with abnormalities. But the fact that a procedure may result in genetic harm to offspring has not deterred clinics from trying a variety of other reproductive technologies.

In other countries, including Britain, a licensing agency determines whether particular clinics should be allowed to offer particular reproductive technologies. In the United States, however, clinics may offer whatever infertility treatments they choose. In fact, one doctor at a fertility clinic told me candidly: "We go from mindside to bedside in two weeks. We make things up and try things on patients. We never get their informed consent, because they just want us to make them pregnant."

Similarly, the United States does not require clinics to report when children created as a result of reproductive technologies are born with abnormalities. This contrasts with Australia, where the government collects data each year on the proportion of children born through *in vitro* fertilization who have genetic abnormalities. In recent months, troubling reports have surfaced about the risks of reproductive technologies. Several studies using the Australian data have shown that children born after intracytoplasmic sperm injection—used when men have low sperm counts—were twice as likely to have major congenital abnormalities as were children conceived naturally.

Another recent report dealt with the risk that embryos may fuse—creating what is called a chimera when multiple embryos created through *in vitro* fertilization are implanted in a woman's womb. In most I.v.F. attempts, more than one embryo is implanted in the womb. In the January 15 issue *of The New England Journal of Medicine*, Lisa Strain, a geneticist at the University of Edinburgh, reported that a chimera had been created through *in vitro* fertilization. Apparently a male embryo and a female embryo put into the mother's womb at the same time fused into a single individual, who had both male and female sex organs at birth. Commentators suggested that other chimeras may have occurred but gone unnoticed because the embryos were of the same sex.

Because the practices of U. S. fertility clinics have been virtually unregulated, it astounded many experts recently when Acting F.D.A. Commissioner Michael Friedman announced that Richard Seed would have to get the agency's approval before attempting to clone humans in the United States. The F.D.A. does have the legal authority to regulate products (such as skin tissue for burn victims) containing cells that have been substantially altered through "more than minimal" manipulation. Some lawyers doubt that this authority covers cloning, however. Even if it does, the guidelines do not require prior approval if a patient's cells are being used for his or her own reproductive purposes. If the F.D.A. can regulate cloning, why hasn't it used the same authority to monitor intracytoplasmic sperm injection, in which DNA (in the form of sperm) is being injected into women's eggs?

The Biotechnology Industry Organization, which represents 750 biotech companies, academic institutions, and state biotechnology centers, is supporting the F.D.A.'s assertion of authority in this field, and is trying to dissuade federal and state lawmakers from enacting moratoriums on human cloning. The organization realizes that if the F.D.A. can regulate human cloning, it is likely to permit cloning once the procedure becomes safer.

The current policy debate over human cloning provides us with the opportunity to clarify what it is that troubles us about human cloning—and to devise laws that reflect those concerns.

Right now, the three cloning bans that are pending in Congress, as well as the 22 bills that have been proposed in state legislatures, vary widely in their scope. Given the latest scientific developments, many of the bills would not even meet their intended goals. A law recently passed in California bars the insertion of nucleic DNA from a human cell into a human egg. But, with the technology developed by First and Dominko using cow eggs, a human egg may no longer be necessary.

What is it, then, that troubles us about human cloning right now? If it is the low potential success rate, then we should be equally concerned about other reproductive technologies, some of which continue to have very low success rates. If our concern is the chance of physical damage to the offspring, we should scrutinize more closely existing reproductive technologies and collect data on the health of the resulting children.

Clones Will Be Psychologically at Risk

I suspect that our concerns run much deeper, though. Many people are worried about the psychological risks to the later-born "twin" created through cloning. Cloning could undermine human dignity by threatening the later child's sense of self and sense of autonomy, which a vast body of research in developmental psychology has demonstrated are important.

The candidates who have been suggested for cloning include a member of an infertile couple, a deceased child, a beloved relative, people with particular favored traits, celebrities, and homosexuals who do not want to reproduce with a member of the opposite sex. But, in attempting to foster in a resulting child the favored traits of a loved one or a celebrity who has been cloned, the adults who raise the child may limit the environmental stimuli to which the child is exposed. The clone of a pianist might not be allowed to spend much time away from the piano, playing baseball or just hanging out with other children. The clone of a dead child might not be offered experiences that the first child had rejected. Clones might wind up in what Francis Pizzulli, a Los Angeles lawyer who has followed developments in biomedicine, describes as a type of "genetic bondage," with improper constraints on their freedom.

If a clone is made of Michael Jordan, and the child breaks his leg at age 10, will his parents consider him worthless? Will he consider himself a failure? If Michael Jordan dies next year of an inheritable cancer, would his young clone become uninsurable? In an era not only of genetic determinism but also of potential genetic discrimination, children saddled with another person's DNA might face psychological and financial risks. Cloning is all too likely to violate what the University of Arizona philosopher Joel Feinberg has called the child's "right to an open future."

If we are concerned about such physical and psychological risks to offspring produced by cloning, the policies that have been proposed thus far are inadequate. Some proposals, such as bills introduced in New York and Illinois, have loopholes because they ban only the creation of a "genetically identical" individual. A human clone would have some mitochondrial DNA from the egg donor, so he or she would not be exactly identical to the individual who provided the rest of the DNA. Thus, an argument could be made that the law would not apply, because the procedure does not create a genetically identical individual.

Other proposals are too specific. Some bills would prohibit the transfer of nucleic material from a somatic cell into a human egg. Not only could such provisions be evaded by using cow eggs, but they also would not prevent cloning via "embryo splitting." An eight-celled embryo can be split into parts to create two or four or eight identical embryos. *In vitro* fertilization clinics eventually want to be able to use this technology to enhance a woman's chances of becoming pregnant.

In fact, the American Society of Reproductive Medicine is working hard to prevent bans on cloning from prohibiting embryo splitting. If all of the embryos resulting from the splitting procedure were implanted at once, the effects would differ from those of cloning, in that all of the siblings would be born at the same time, and each would have an open future. But a woman with eight identical embryos might not want to risk her health or that of the offspring by implanting them all at once, so she might freeze some of the embryos to be implanted later. And that approach, as is the case with cloning from an existing individual, would create the psychological risks of later-born twins.

In fact, when I first heard Richard Seed speak, at an international meeting on *in vitro* fertilization in 1980, he advocated "splitting an embryo in half, freezing half, growing up the first half to see if it got into Harvard, and then deciding to implant the second twin."

Current U.S. Regulation of Reproduction Is Insufficent

It is clear to me that neither the regulatory approach of the F.D.A. nor the narrow bans proposed by Congress and state legislatures are sufficient. I believe that, in line with the British model, we need to create a governmental oversight body with the authority to license fertility clinics, assess what reproductive technologies may be safely offered and by whom, and require the, collection of follow-up data on the children created by these technologies.

We also need a process in this country for debating the advisability of particular reproductive services. When various procedures first were proposed in Canada, a Royal Commission on Reproductive and Genetic Technologies, which included experts from many disciplines, spent two years assessing Canada's cultural values in a range of ways, from anthropological studies to public responses on a toll-free number. The commission ultimately recommended bans on cloning, on paid surrogate motherhood, and on techniques that would allow parents to select the sex of their child. It said Canadians' values opposed the "commodification" of people.

In making national policy about new means of creating children in the United States, we don't appear to have similar cultural values. The American attitude is one of "show me the money, and the technology will become available somehow." Now that the Raelians have shown Seed the money, though, we need to give serious thought to whether our laissez-faire market mechanisms are the best determinants of how children should be brought into the world.

VIII

Is Sex Obsolete?

Introduction to Part VIII

Cloning is asexual reproduction. The beginning of human cloning therefore raises the question of the function of sex in human life. The theory that a large population of genetically identical individuals would be too easy a prey for invading disease organisms can be interpreted as an argument against proceeding too far with cloning, or as the prediction that cloning will not be able to get very far. Both these interpretations can be detected in the following readings.

Reading 39 by David Stipp concisely explains a leading theory of why sex evolved and why it is still necessary. Essentially the same theory is outlined by *The New Leader*, in Reading 40. Finally, Michael Mautner argues (Reading 41) that the human species still needs to evolve further, and that biotechnology issues should be settled by an informed "biodemocracy."

39

Sex Is Still Necessary

David Stipp

To put the recent cloning of a sheep in perspective, it helps to keep in mind two things: dogs and sex.

The British scientists' achievement ranks as a major first. They showed that DNA from an adult mammal can revert to an embryonic state and then duplicate the animal whence it came. But as a feat of genetic manipulation, Dolly's cloning hardly compares with what dog breeders have done over the centuries: mold a mammal's genome like play-doh to create a menagerie worthy of Dr. Seuss. Imagine what the bioengineers who fashioned those wadded-up hippo-faced canines, the Shar-Pei breed, might have done with Dolly's kind.

We'll get to sex by way of a question: Will Dolly the sheep change our lives? For all the furor about cloning, it has inspired few ideas for truly doable things—the kind, say, that would play in a Peoria investors' club.

Consider cloning herds of supercows from a champion milk producer. Setting aside the fact that dairy farms are constantly going bust because we're awash in milk, the idea is nevertheless deeply flawed. Biologist Ursula Goodenough came close to the reason with her quip to the *New York Times*: If cloning were perfected, she said, "there'd be no need for men." Very funny, Ursula—we all know men, women, and whoopee were meant to be. Sex isn't just fun, it's serious business.

Here's why sex really matters—and why cloning makes for bad breeding. Our microbial enemies constantly evolve ways to defeat our defenses and invade our cells. Sex is our brilliant countermeasure, originated by evolution eons ago. By mingling genes, male and female creatures arm their offspring with novel DNA combinations. Microbes equipped to burglarize one generation's cells find the cellular locks changed in the next. In

---

This appeared as "The Real Biotech Revolution" in *Fortune*, v135 (March 31, 1997), pp. 54–5.

short, without sex, we'd soon be toast for germs. And cloned, genetically identical cows would be sitting ducks for epidemics.

Another proposal, cloning gene-tweaked animals that make human medicines in their milk, has more merit. It would keep the beasts' precious genomes from being sullied over time by sexual gene mixing. But Dolly was the only success in 277 cloning tries. There seem to be better ways already to generate living drug factories. This month, for example, researchers reported making human hemoglobin—needed for blood substitutes—in bioengineered tobacco plants, of all things.

Of course, the excitement about cloning has little to do with its practical value. What really rivets us is the idea of cloning ourselves. Indeed, Dolly represents the perfect discovery for the me Generation—it can now fancy itself becoming the Me Me Me Generation. We'll not bore you with a litany of dark possibilities—surely you'll scream if you hear the word Orwellian again.

Instead, a modest prediction: The benefits of human cloning, if any, won't be compelling enough to induce scientists to risk their careers, and possibly even their necks, trying to realize them.

But they'll risk much to understand how genes work. That knowledge will pay in ways cloning never can. Smart money in places like Peoria already is flowing to the practitioners of a new branch of biotech, genomics, which promises to extend the genetics revolution far beyond mere therapies for our bodily scourges—to actually curing and preventing them. Recently Silicon Valley has emerged as the Florence of this Renaissance, thanks largely to biochips in the works there that promise to be the X-ray machines of the gene age.

So read on and give Dolly a rest. She really deserves it.

40

Males Have at Least One Use

The New Leader

With the announcement of the successful cloning of an adult mammal the other week, the thoughts of pundits like George Will turned to human dignity, the soul, and the possibility of immortality. Ours turned to sex. What, we wondered, is the point of sexual congress anymore?

Ahem, one might respond, if you have to ask ... But seriously. Assuming that human cloning will be possible within the next decade—a fairly cautious forecast—might not sex become obsolete? Biologically, it is far more efficient to reproduce by turning out perfect genetic replicas of oneself, the way dandelions do; or by virgin birth, the way greenflies do; or by fissioning, the way amoebas do. And might not males become obsolete, or even extinct, as some of the barmier feminists profess to hope? Men are a waste: They do not gestate babies. Better that humans should be hermaphrodites; or have 10,000 different genders, like the toadstool—then, unless you happen to be a same-sexer, almost everyone you bumped into would be a potential partner.

Sex and reproduction first became disjoined in the late 19th century, with the invention of reliable contraception. Today, with artificial insemination and *in vitro* fertilization, sex is neither sufficient nor necessary for baby-making. Once cloning becomes an option, moreover, one won't need to mix up one's chromosomes with those of another member of *Homo sapiens*; they can be perpetuated perfectly intact.

Of course, it is only odious narcissists who will actually do this (why does poor Donald Trump always come to mind?). Yet according to Darwinian psychology, we should all want to do it. Well, suppose we all did do it. What would be the consequences for the species were sex to become just a recreational pastime, utterly divorced from procreation?

This appeared as "Between Issues" in *The New Leader*. v80 (February 24 1997). p. 2.

question only in the last couple of decades. Before then, sex was, scientifically speaking, an enigma. True, the sexual mixing of genes speeds up evolution. There is, however, no teleological imperative for a species to evolve; some kinds of fish have not changed a bit over the last million years. If certain members of a given species were to start making copies of themselves asexually, they would quickly come to predominate; because, as one evolutionary biologist put it, "asex is twice as fecund as sex."

But soon they would be in trouble—for a reason that brings us, finally, to the point of sex. Ironic as it may seem in the present age, sex evolved because of its value in thwarting disease. The sexual mixing of genes, the evidence indicates, is analogous to changing the locks constantly against invading microbes, which are themselves mixing their genes to find the new key. Sex, in other words, is a zero-sum game that organisms play against parasites over the generations. (Perhaps that is why, in the old Latin proverb, "every animal is sad after intercourse.").

Thus, though we have recently seen the End of History, of Art and of Nature, the End of Sex had better not be imminent. It is comforting to know that males are not useless after all.

41

Cloning Puts Future Human Evolution in Doubt

Michael Mautner

The recent cloning of the first mammal brings the prospects of human cloning closer to reality. Now the public should ponder the implications. Among these, the most important is the effect on our future evolution.

Cloning will be attractive because of some medical uses. Genetic replicas of geniuses might also benefit society. On the other hand, ruthless and egocentric despots may replicate themselves millions of times over. Cloning on a large scale would also reduce biological diversity, and the entire human species could be wiped out by some new epidemic to which a genetically uniform population was susceptible.

Beyond these important but obvious results, cloning raises problems that go to the core of human existence and purpose. One important fact to recognize is that cloning is asexual reproduction. It therefore bypasses both the biological benefits of normal reproduction and the emotional, psychological, and social aspects that surround it: courtship, love, marriage, family structure. Even more importantly, if cloning became the main mode of reproduction, human evolution would stop in its tracks.

In sexual reproduction, some of the genetic material from each parent undergoes mutations that can lead to entirely new biological properties. Vast numbers of individual combinations become possible, and the requirements of survival—and choices of partners by the opposite sex—then gradually select which features will be passed on to the following generations.

Cloning will, in contrast, reproduce the same genetic makeup of an existing individual. There is no room for new traits

This appeared as "Will Cloning End Human Evolution?" in *The Futurist*, v31 (November/December 1997), p. 68. Published by the World Future Society, 7910 Woodmont Avenue, Suite 450, Bethesda, Maryland 20814.

to arise by mutation and no room for desirable features to compete and win by an appeal to the judgment of the opposite sex. The result: Human evolution is halted.

Is it necessary for the human species to evolve further? Absolutely! We are certainly far from achieving perfection. We are prone to diseases, and the capacity of our intelligence is limited. Most importantly, human survival will depend on our ability to adapt to environments beyond Earth—that is, in the rich new worlds of outer space.

Some people question whether we can save ourselves from man-made environmental disasters on Earth, whose resources are already pressured by human population growth. And limiting the population to one planet puts us at risk of extinction from all-out nuclear or biological warfare, climate change, and catastrophic meteorite impacts.

Humanity could vastly expand its chances for survival by moving into space, where we would encounter worlds with diverse environments. To live in space, we will have to increase our tolerance to radiation, to extremes of heat and cold, and to vacuum. We will also need more intelligence to construct habitats. Our social skills will need to advance so that billions of humans can work together in the grand projects that will be needed.

If we are to expand into space, we surely cannot freeze human evolution. The natural (and possibly designed) mechanisms of evolution must therefore be allowed to continue.

Socially, the relations between the sexes underlie most aspects of human behavior. The rituals of dating, mating, and marriage and the family structures that surround sexual reproduction are the most basic emotional and social factors that make our lives human. Without the satisfactions of love and sex, of dating and of families, will cloned generations even care to propagate further?

Cloning therefore raises fundamental questions about the human future: Have we arrived yet at perfection? Where should we aim future human evolution? What is the ultimate human purpose? The prospect of human cloning means that these once-philosophical questions have become urgent practical issues.

As living beings, our primary human purpose is to safeguard, propagate, and advance life. This objective must guide our ethical judgments, including those on cloning.

Our best guide to this purpose is the love of life common to most humans, which is therefore reflected in our communal judgment. All individuals who sustain the present and build the future should have the right to participate equally in these basic decisions. Our shared future may be best secured by the practice of debating and voting on such biotechnology issues in an informed "biodemocracy."

IX

The Nightmare of the Super Race

Introduction to Part IX

Ever since the western world was plunged into mass destruction by Hitler's doctrine of racial superiority, people have been alert to the possibility of the harm that can arise from attempts to breed a better human stock. Immediately following Dolly came renewed speculation about the potential application of cloning for neo-Hitlerian ends. Several of the earlier readings in this book briefly mention this concern, which had been popularized before Dolly by Ira Levin's novel, *The Boys from Brazil.*

Barry Came's article (Reading 42) links Dolly with Hitler. Brian Johnson illustrates (Reading 43) the uses, sometimes totalitarian, cloning has found in popular culture. Jeffrey Kluger (Reading 44) conjures up several future forms of demand for cloning, including the aging dictator eager to secure endless self-replication.

42

We Are Psychologically Unprepared for Cloning

Barry Came

In Harold Shapiro, U.S. President Bill Clinton may have found the perfect candidate to explore the mysteries surrounding animal—and potentially human—cloning. To be sure, as president of Princeton University in New Jersey, the 61-year-old Montreal-born economist is a respected academic. But Shapiro brings another fitting credential to his assignment as chairman of the Federal Bioethics Advisory Commission, charged last week with preparing a report for Clinton on the legal and ethical implications of the new cloning technology. He and his brother, Bernard, also a university administrator—at McGill in Montreal—are twins. "So I guess you could say that I was specially made for this job," says Harold with a chuckle.

Unlike many of his academic colleagues, Shapiro is not unduly concerned about the ramifications likely to flow from the creation of that sheep called Dolly. "I have to admit it's a startling event that poses a host of questions," he acknowledges. "But at the same time, I have every confidence that we'll be able to do something to keep it under control." Modern society has learned to deal safely with "much scarier" technology, he argues, citing nuclear weapons and poison gas as examples. "The chances are," says Shapiro, "this entire affair is going to end up producing a lot more benefits than costs."

That is certainly the view of pharmaceutical firms involved in the effort to transform barnyards of domestic animal species into what amounts to four-legged drug factories. They see Dolly as merely another step along the road towards having genetically altered sheep, cows and pigs produce not only more and better milk and meat but also human proteins for use in the fight against cancer, cystic fibrosis and other diseases. They also talk of using

---

This appeared as "The Prospect of Evil" in *Maclean's*, v110 (March 10 1997), p. 59.

cloning for the wholesale production of spare body parts—replacement hearts and livers, lungs and kidneys.

But the shock of seeing such a radical technology come so close to human application raises troubling questions about its possible uses. It is frightening how easily people can be subverted to evil purposes, says Dr. Gerald Klassen, a bioethicist and a professor of medicine at Dalhousie University in Halifax. "We have the idea that doctors are particularly ethical and that they will always make the right choices," says Klassen. "But then you look at the extraordinarily high participation rate of the medical profession in the eugenics experiments of Nazi Germany."

While many European countries have regulations outlawing experimentation with human cloning, there are no such laws yet in Canada or the United States. Last June, the Liberal government in Ottawa introduced legislation that would prohibit human cloning, along with some other reproductive techniques. Based on the recommendations of the 1993 report of the Royal Commission on New Reproductive Technologies, the bill goes before House of Commons committee hearings this month, but is unlikely to become law before an election is called. "It's hard to see any ethically defensible use of cloning for human beings," says University of British Columbia geneticist Patricia Baird, who wrote the commission report. She draws an important distinction between cloning and earlier, more acceptable, technologies like test-tube reproduction and *in vitro* fertilization. "A baby born *in vitro* would have an egg and a sperm from her mother and father," she notes. "With cloning, you simply copy one of the cells of an adult person."

Margaret Somerville of McGill's Centre for Medicine, Ethics and the Law counts cloning as the third miracle of modern medical science, after heart transplant surgery and test-tube reproduction. But she, too, wants it prohibited for humans. Somerville attributes revulsion to the very notion of cloning to "a moral intuition, an innate gut reaction that we've got to listen to when we sit down and do our cool logic." What she sees developing is an argument about the essence of humanity. "It is a radical shift in the whole nature of the uniqueness of each human from a genetic point of view," says Somerville.

So radical, in fact, that human beings are "psychologically unprepared" for the entire concept of cloning, says psychologist Charles Crawford of Simon Fraser University in Burnaby, B.C.

"When we get the possibility of several individuals who are identical, it is hard for us to know how to react, to know whether it is good or bad." But now that it has been accomplished with sheep, he expects it to happen with humans. "We will see people set up cloning clinics in less regulated, more enterprising underdeveloped countries," says Crawford. "It could cause a major change in male-female relationships. We could totally disengage reproduction from sexuality."

That, of course, is a concern from a theological perspective. "Procreation without the sexual act of husband and wife is considered morally wrong," notes Most Rev. Adam Exner, Roman Catholic archbishop of Vancouver. He also foresees frightening social consequences resulting from dehumanizing the act of procreation. "For example, what kind of a self-image would a cloned offspring have?" wonders Exner. "What psychological problems might that cause? How would a lot of cloned individuals change society?" If Dolly the cloned sheep is a harbinger of what lies ahead, it may be time to reach for answers to those hard questions.

43

Self-Replication Is Big in Hollywood

Brian D. Johnson

Now that science has figured out how to Xerox sheep, fears are running rampant that a brave new breed of human replicants is just around the corner. Big deal. Hollywood, where they have been treating people like sheep for ages, has already perfected the cloning business. Take, for example, the movie about the mad scientist who fabricates his offspring in the laboratory. Which movie? Take your pick. Ever since the early Frankenstein flicks, the premise itself has been cloned over and over and over again.

The classic example of a gene-cloning movie is the 1978 thriller *The Boys from Brazil*, in which Gregory Peck is ludicrously miscast as Dr. Josef Mengele, a Nazi psychopath who incubates a brood of spoiled teenage boys with blue eyes and shocks of straight black hair from samples of skin and blood donated by Adolf Hitler before his death. Mengele's goal is to resurrect the Third Reich with a Hitler clone. Laurence Olivier plays Lieberman, the Nazi hunter who tracks down the bad doctor. In a scene designed to lend the premise some plausibility, he listens to a scientist—a sane one—explain how he has cloned rabbits by injecting donor cells into eggs stripped of their genetic nucleus.

"And this can be done with humans?" asks the incredulous Lieberman. "It's monstrous!" The scientist gives him a blank look. "Why?" he asks. "Wouldn't you want to live in a world of Mozarts and Picassos? Of course it's only a dream. Not only would you have to reproduce the genetic code of the donor, but the environmental background as well."

Good thinking. Not every cloning flick is so fastidious—certainly not Woody Allen's *Sleeper* (1973), in which a dictator's disembodied nose held the blueprint for cloning a new world

---

This article appeared as "Body Doubles" in *Maclean's*, v110 (March 10 1997), pp. 56–7.

business, leading to the propagation of a runaway species that could consume the planet. In the Frankenstein tradition, however, the scientists can have benign intentions. The Nobel laureate played by Peter O'Toole in *Creator* (1985) was just trying to make a facsimile of his dead wife to keep him company in his old age. And the dinosaur-incubating eggheads in *Jurassic Park*—which will be cloned in a blockbuster sequel this summer—just wanted to build a zoo.

There is, of course, an entire genre of sci-fi movies populated by cybernetic clones—ranging from the replicants in *Blade Runner* to the Arnold Schwarzenegger gladiator in *Terminator 2*. They are, basically, disposable people. Their flesh can be graphically gored, squished, punctured, cut and ripped, and for some reason the violence is supposed to seem less offensive because it is just machine flesh. Aliens, meanwhile, are old hands at cloning people as hosts. Movies ranging from *Invasion of the Body Snatchers* (1956) to *Species* (1995) have taught us that soulless mutants may be lurking under the skin of even the most regular-looking folks.

Finally, there is the clone as household drone. In last year's comedy *Multiplicity*, Michael Keaton played an overworked construction foreman who replicates himself with the help of a local geneticist. The first clone is a macho overachiever, the second an effeminate New Man, and the third (a clone of a clone) a blithering idiot. The film fared poorly at the box office, putting a damper on future projects in the same vein. But Dolly's fame has changed all that—at the studios, cloning scripts are suddenly in demand.

But then cloning is what Hollywood is all about—call it the dubbing-down of mass culture. The movie industry thrives on duplicating the same formulas again and again, turning out sequels, remakes and knock-offs from an ever-shrinking gene pool of creativity. It also clones its stars, replicating a single personality into common currency. Andy Warhol, pop art's genius freak, figured it out in the 1960s with his serial silk screens of such icons as Marilyn Monroe, Elvis Presley and the Mona Lisa. Duplicating both the living and the dead, Warhol cloned the very tissue of celebrity, printing it up like sheets of counterfeit money. He was both mocking and minting Hollywood culture, and in case anyone missed the point, he also ran off silk-screened images of dollar bills.

In a globalized culture where a face can become an instant franchise—witness the proliferation of merchandise emblazoned with the Edvard Munch painting *The Scream*—cloning is a metaphor that has come full circle. In the typical horror fantasy, what draws the laboratory scientist to play God is a lust for immortality. In their own way, artists have always hoped to leave a mark that lasts beyond their death. But Hollywood has industrialized that desire. And if special-effects wizards have their way, the stars of the future could actually be clones—the digital kind. Director Robert Zemeckis, who brought John F. Kennedy back to life in *Forrest Gump*, has predicted that actors may soon be able to extend their careers indefinitely by banking computer clones of themselves in their prime. They could co-star with their younger selves in flashbacks, and be cast in movies long after their death.

But some stars might prefer to replicate themselves in the flesh. Madonna, for instance, could sidestep that whole messy business of finding a mate. She could conceive her own immaculate bundle of blond ambition—just like a virgin.

44

The Ethics of Cloning May Prove Harder than the Science

Jeffrey Kluger

It's a busy morning in the cloning laboratory of the big-city hospital. As always, the list of people seeking the lab's services is a long one—and, as always, it's a varied one. Over here are the Midwestern parents who have flown in specially to see if the lab can make them an exact copy of their six-year-old daughter, recently found to be suffering from leukemia so aggressive that only a bone-marrow transplant can save her. The problem is finding a compatible donor. If, by reproductive happenstance, the girl had been born an identical twin, her matching sister could have produced all the marrow she needed. But nature didn't provide her with a twin, and now the cloning lab will try. In nine months, the parents, who face the very likely prospect of losing the one daughter they have, could find themselves raising two of her—the second created expressly to help keep the first alive.

Just a week after Scottish embryologists announced that they had succeeded in cloning a sheep from a single adult cell, both the genetics community and the world at large are coming to an unsettling realization: the science is the easy part. It's not that the breakthrough wasn't decades in the making. It's just that once it was complete—once you figured out how to transfer the genetic schematics from an adult cell into a living ovum and keep the fragile embryo alive throughout gestation—most of your basic biological work was finished. The social and philosophical temblors it triggers, however, have merely begun.

Only now, as the news of Dolly, the sublimely oblivious sheep, becomes part of the cultural debate, are we beginning to come to terms with those soulquakes. How will the new technology be regulated? What does the sudden ability to make

---

This appeared as "Will We Follow the Sheep?" in *Time*. v149 (March 10 1997), pp. 66–7.

genetic stencils of ourselves say about the concept of individuality? Do the ants and bees and Maoist Chinese have it right? Is a species simply an uberorganism, a collection of multicellular parts to be die-cast as needed? Or is there something about the individual that is lost when the mystical act of conceiving a person becomes standardized into a mere act of photocopying one?

Last week President Clinton took the first tentative step toward answering these questions, charging a federal commission with the task of investigating the legal and ethical implications of the new technology and reporting back to him with their findings within 90 days. Later this week the House subcommittee on basic research will hold a hearing to address the same issues. The probable tone of those sessions was established last week when Harold Varmus, director of the National Institutes of Health (NIH), told another subcommittee that cloning a person is "repugnant to the American public."

Though the official responses were predictable—and even laudable—they may have missed the larger point. The public may welcome ways a government can regulate cloning, but what's needed even more is ways a thinking species can ethically fathom it. "This is not going to end in 90 days," says Princeton University president Harold Shapiro, chairman of President Clinton's committee. "Now that we have this technology, we have some hard thinking ahead of us."

Also waiting in the cloning lab this morning is the local industrialist. Unlike the Midwestern parents, he does not have a sick child to worry about; indeed, he has never especially cared for children. Lately, however, he has begun to feel different. With a little help from the cloning lab, he now has the opportunity to have a son who would bear not just his name and his nose and the color of his hair but every scrap of genetic coding that makes him what he is. Now that appeals to the local industrialist. In fact, if this first boy works out, he might even make a few more.

Of all the reasons for using the new technology, pure ego raises the most hackles. It's one thing to want to be remembered after you are gone; it's quite another to manufacture a living monument to ensure that you are. Some observers claim to be shocked that anyone would contemplate such a thing. But that's naive—and even disingenuous. It's obvious that a lot of people would be eager to clone themselves.

"It's a horrendous crime to make a Xerox of someone,"

argues author and science critic Jeremy Rifkin. "You're putting a human into a genetic straitjacket. For the first time, we've taken the principles of industrial design—quality control, predictability—and applied them to a human being."

But is it really the first time? Is cloning all that different from genetically engineering an embryo to eliminate a genetic disease like cystic fibrosis? Is it so far removed from *in vitro* fertilization? In both those cases, after all, an undeniable reductiveness is going on, a shriveling of the complexity of the human body to the certainty of a single cell in a Petri dish. If we accept this kind of tinkering, can't we accept cloning? Harvard neurobiologist Lisa Geller admits that intellectually, she doesn't see a difference between *in vitro* technology and cloning. "But," she adds, "I admit it makes my stomach feel nervous."

More palatable than the ego clone to some bioethicists is the medical clone, a baby created to provide transplant material for the original. Nobody advocates harvesting a one-of-a-kind organ like a heart from the new child—an act that would amount to creating the clone just to kill it. But it's hard to argue against the idea of a family's loving a child so much that it will happily raise another, identical child so that one of its kidneys or a bit of its marrow might allow the first to live. "The reasons for opposing this are not easy to argue," says John Fletcher, former ethicist for the NIH.

The problem is that once you start shading the cloning question—giving an ethical O.K. to one hypothetical and a thumbs-down to another—you begin making the sort of *ad hoc* hash of things the Supreme Court does when it tries to define pornography. Suppose you could show that the baby who was created to provide marrow for her sister would forever be treated like a second-class sibling—well cared for, perhaps, but not well loved. Do you prohibit the family from cloning the first daughter, accepting the fact that you may be condemning her to die? Richard McCormick, a Jesuit priest and professor of Christian ethics at the University of Notre Dame, answers such questions simply and honestly when he says, "I can't think of a morally acceptable reason to clone a human being."

In a culture in which not everyone sees things so straightforwardly, however, some ethical accommodation is going to have to be reached. How it will be done is anything but clear. "Science is close to crossing some horrendous boundaries," says

Leon Kass, professor of social thought at the University of Chicago. "Here is an opportunity for human beings to decide if we're simply going to stand in the path of the technological steamroller or take control and help guide its direction."

Following the local industrialist on the appointments list is the physics laureate. He is terminally ill. When he dies, one of the most remarkable minds in science will die with him. Reproductive chance might one day produce another scientist just as gifted, but there is no telling when. The physics laureate does not like that kind of uncertainty. He has come to the cloning lab today to see if he can't do something about it.

If the human gene pool can be seen as a sort of species-wide natural resource, it's only sensible for the rarest of those genes to be husbanded most carefully, preserved so that every generation may enjoy their benefits. Even the most ardent egalitarians would find it hard to object to an Einstein appearing every 50 years or a Chopin every century. It would be better still if we could be guaranteed not just an Einstein but *the* Einstein. If a scientific method were developed so that the man who explained general relativity in the first half of the century could be brought back to crack the secrets of naked singularities in the second, could we resist using it? And suppose the person being replicated were researching not just abstruse questions of physics but pressing questions of medicine. Given the chance to bring back Jonas Salk, would it be moral not to try?

Surprisingly, scientific ethicists seem to say yes. "Choosing personal characteristics as if they were options on a car is an invitation to misadventure," says John Paris, professor of bioethics at Boston College. "It is in the diversity of our population that we find interest and enthusiasm."

Predetermining Personal Characteristics Would Be Prone to Fads

Complicating things further, the traits a culture values most are not fixed. If cloning had existed a few centuries ago, men with strong backs and women with broad pelvises would have been the first ones society would have wanted to reproduce. During the industrial age, however, brainpower began to count for more than muscle power. Presumably the custodians of cloning technology at that historical juncture would have faced the prospect of letting previous generations of strapping men and fecund women die out and replacing them with a new population of intellectual giants. "What is a better human being?" asks Boston University ethicist George Annas. "A lot of it is just fad."

Even if we could agree on which individuals would serve as humanity's templates of perfection, there's no guarantee that successive copies would be everything the originals were. Innate genius isn't always so innate, after all, coming to nothing if the person born with the potential for excellence doesn't find the right environment and blossom in it. A scientific genius who's beaten as a child might become a mad genius. An artist who's introduced to alcohol when he's young might merely become a drunk. A thousand track switches have to click in sequence for the child who starts out toward greatness to wind up there. If a single one clicks wrong, the high-speed rush toward a Nobel Prize can dead-end in a makeshift shack in the Montana woods. Says Rabbi Moshe Tendler, professor of both biology and biblical law at Yeshiva University in New York City: "I can make myself an Albert Einstein, and he may turn out to be a drug addict."

The despot will not be coming to the cloning lab today. Before long, he knows, the lab's science will come to him—and not a moment too soon. The despot has ruled his little country for 30 years, but now he's getting old and will have to pass his power on. That makes him nervous; he's seen what can happen to a cult of personality if too weak a personality takes over. Happily, in his country that's not a danger. As soon as the technology of the cloning lab goes global—as it inevitably must—his people can be assured of his leadership long after he's gone.

This is the ultimate nightmare scenario. The Pharaohs built their pyramids, the Emperors built Rome, and Napoleon built his Arc de Triomphe—all, at least in part, to make the permanence of stone compensate for the impermanence of the flesh. But big buildings and big tombs would be a poor second choice if the flesh could be made to go on forever. Now, it appears, it can.

The idea of a dictator's being genetically duplicated is not new—not in pop culture, anyhow. In Ira Levin's 1976 book *The Boys from Brazil* a zealous ex-Nazi bred a generation of literal Hitler Youth—boys cloned from cells left behind by the Fuhrer. Woody Allen dealt with a similar premise a lot more playfully in his 1973 film *Sleeper*, in which a futuristic tyrant is killed by a bomb blast, leaving nothing behind but his nose—a nose that his followers hope to clone into a new leader. Even as the fiction of one decade becomes the technology of another, it's inevitable that this technology will be used—often by the wrong people.

"I don't see how you can stop these things," says bioethicist Daniel Callahan of the Hastings Center in Briarcliff Manor, New York. "We are at the mercy of these technological developments. Once they're here, it's hard to turn back."

Hard, perhaps, but not impossible. If anything will prevent human cloning—whether of dictator, industrialist or baby daughter—from becoming a reality, it's that science may not be able to clear the ethical high bar that would allow basic research to get under way in the first place. Cutting, coring and electrically jolting a sheep embryo is a huge moral distance from doing the same to a human embryo. It took 277 trials and errors to produce Dolly the sheep, creating a cellular body count that would look like sheer carnage if the cells were human. "Human beings ought never to be used as experimental subjects," Shapiro says simply.

Whether they will or not is impossible to say. Even if governments ban human cloning outright, it will not be so easy to police what goes on in private laboratories that don't receive public money—or in pirate ones offshore. Years ago, Scottish scientists studying *in vitro* fertilization were subjected to such intense criticism that they took their work underground, continuing it in seclusion until they had the technology perfected. Presumably, human-cloning researchers could also do their work on the sly, emerging only when they succeed.

Scientists don't pretend to know when that will happen, but some science observers fear it will be soon. The first infant clone could come squalling into the world within seven years according to Arthur Caplan, director of the Center for Bioethics at the University of Pennsylvania. If he's right, science had better get its ethical house in order quickly. In calendar terms, seven years from now is a good way off; in scientific terms, it's tomorrow afternoon.

X

Science versus the Public Good?

Introduction to Part X

In Reading 45, Art Levinson suggests that, while the prohibition of the cloning of an entire human being is a legitimate political choice, attempts to craft such legislation could be in danger of hurting valuable research. In his consideration of the role of public interest in regulating science, Stephen Toulmin (Reading 46) holds both that the First Amendment does not protect science from public supervision and that recombinant DNA research is not inherently "forbidden knowledge." Jonathan Beckwith (Reading 47) believes that we should seriously consider whether recombinant DNA research should be pursued at all, given its potential for abuse, especially by the powerful against the less powerful groups in society.

45

Government Should Be Careful to Avoid Stopping Legitimate Research

Art Levinson

The biotechnology and research community has been very open and public about its support for the President's request for a voluntary moratorium on activities that could lead to the cloning of entire human beings. This exercise of responsibility in science is consistent with our long history of restraint in the pursuit of basic biomedical research. We do not plan or seek to clone entire human beings. In addition, we fully recognize the existence of various federal laws setting out the jurisdiction of the Food and Drug Administration which, when taken together, would bar the commercialization of cloning of entire human beings. Because of this moratorium and existing legal limitations on action, it is possible to deliberate and exercise caution and restraint in legislating this issue.

The reality of modern biomedical research is that it is difficult to predict in advance exactly how specific, even esoteric, areas of research will produce breakthroughs. As Michael Bishop (cancer researcher, Nobel laureate in medicine and my colleague from the University of California, San Francisco) spoke of this issue recently, in 1968 his work with Dr. Harold Varmus, and Professor Herb Boyer would have never been foreseen as leading to breakthroughs in recombinant DNA research and cancer genetics. Similarly, work done in the 1980s on transgenic animals by Dr. Phil Leder, of Harvard, and others, would not have easily been understood as being essential to the development of animal models that could facilitate dramatic advances in our ability to test new AIDS therapies.

It is also the case that with virtually every scientific advance there are voices that seek to delay legitimate, if misunderstood, advances in science. In the early 1970s, some government officials sought to vary virtually all recombinant DNA research out of exaggerated fears about the safety of the technology. Researchers

---

This appeared in the *Congressional Record* of February 10, 1998 p. S578–579.

and companies voluntarily adopted a moratorium on some research until more information was obtained. Fortunately, the calls for more radical local or federal regulation were rejected. The self-regulatory efforts by industry and the research community worked, and there were no significant safety issues to arise out of that research.

In the 1980s some critics advocated bans on transgenic animal research out of fear of science. These requests for a halt to research were often based on assertions of pseudo-science. Again, we are fortunate that Congress did not act to bar the creation of transgenic animals, which are now so commonly used in drug development, especially in AIDS research. In addition, transgenic animals may someday be used for the actual production of pharmaceutical compounds. This hope for pure protein production at a lower cost is yet to be realized, but if Congress had acted in the 1980s to end research, patients would have had that hope foreclosed.

Now Congress is faced with difficult decisions about how to react to a single experiment in sheep. ... Yet, determining how to prohibit the act of cloning an entire human being has proven to be a daunting task. ...

Most importantly, in considering restrictions on scientific research in the private sector (as opposed to previously enacted limitations on the expenditure of federal funds), great care must be exercised. In addition to the legal rights of persons to free expression and inquiry in the private market, there is little precedent for imposing limitations on research except for reasons of safety or other narrowly crafted circumstances.

In this instance, there are multiple possibilities of promising research with somatic cells. Our hope in the research community is that this branch of research will lead to discoveries that permit us to develop new cures and treatments for serious and unmet medical needs. Some of our colleagues in academe have already begun exploring questions of how to turn on and off these somatic cells so that new biological material could be generated for transplantation and for other therapeutic purposes. At this point in the discovery process, it is not known exactly how to accomplish this therapeutic goal, but one possible way is to use the technique known as somatic nuclear cell transfer. ...

There seems to be little dispute within the Congress about the current inappropriateness of using somatic nuclear cell transfer technology to create an embryo which is implanted into the uterus, with the goal being reproductive in nature. On the other hand, it is hard to understand why scientists should become criminals if they pursue legitimate new therapies for heart disease, cancer, diabetes, and other diseases, and if their research has no prospect or intent of creating an entire cloned human being.

Given our current state of knowledge, there is no reasonable prospect for creating a new human being unless an embryo is implanted into the uterus of a woman. Thus, the approach should be to adopt a bill that effectively bars what the political consensus wants to prohibit, while simultaneously retaining the option of research that is aimed at new therapies, not at reproductive ends. ...

If criminal penalties or asset forfeiture are threatened for research activities, there is likely to be a chilling effect on research in this entire area. Moreover, there are additional sanctions available under the Food, Drug and Cosmetic Act to address human cloning.

The work done by existing entities, such as the Recombinant DNA Advisory Committee of the NIH, and the NIH-DOE Working Group on Ethical, Legal, and Social Implications of Human Genome Research, has advanced the public discussion. In this regard, the work already done by the President's Commission on the topic of cloning entire human beings has materially assisted the national debate on this topic. We leave to the political process questions of whether any such bioethics commission should be situated in the Executive Branch and who should exercise the appointment authority.

There are several caveats worth noting, however:

• Past history, here and in Europe, suggests that there is a real risk that any such commission could inadvertently begin to function as a new regulatory entity and serve to delay the approval of new treatments for patients. This temptation should be avoided at all costs by explicitly limiting the role of the commission.

• There is a risk that any new commission will be led by other political agendas into discussions that do not advance progress on improving human health. This temptation should also be avoided by narrowly circumscribing the commission's charter.

• The composition of any commission should broadly reflect the best available thinking in science, law, and ethics. The mere prohibition on political officials serving on such a panel is not likely sufficient to prevent the politicization of the appointment process. There are, I understand, precedents that permit certain relevant professional societies to offer lists of nominees to an appointing authority. This approach would appear to mitigate the risk of an overly political appointment process.

... Can we simultaneously advance science and the search for cures for serious diseases while also barring the cloning of entire human beings? We believe that to foster further dialogue and deliberation can help achieve that common goal.

46

Recombinant DNA Research Is Not Forbidden Knowledge

Stephen E. Toulmin

To begin with the question of responsibility for risks, I think one can say quite flatly that if there is any serious possibility of recombinant DNA research leading to harm among the general public, then the right of the public authorities to intervene is beyond dispute. Of course, I realize that there is a lot of disagreement among well-qualified scientists about the scale and nature of any such possible risks. My own primary training was in physics rather than biochemistry, and I have to leave it to others to settle that matter among themselves. It is up to those who understand the issues much better than I ever could to arrive at some sort of consensus first, and they must not be surprised if everyone else adopts a cautious and conservative attitude toward the matter in the meanwhile.

There Is No Unqualified Right to Freedom of Research

But there is one extreme view of the issue which it seems to me can be dismissed out of hand. Some of my scientific friends are so shocked by the spectacle of the Cambridge City Council placing hurdles in the way of Harvard's recombinant DNA program that they have reacted by making some quite indefensible constitutional claims. They have suggested that scientific research is not merely a constitutionally protected activity: under the Bill of Rights—that is, that the right to do whatever scientific research one thinks fit without state interference is guaranteed by the first amendment—but that the unimpeded exercise of that right is protected absolutely. As to that there are two fairly brief things which can be said.

First, it is not certain that the first amendment does in fact

This appeared as "The Research and the Public Interest" in *Research with Recombinant DNA,* 1977. National Acadamy of Sciences, Washington D.C. pp. 101–106.

cover the right to do whatever scientific research one thinks fit (or, indeed whether it covers the right to do scientific research at all): the matter has never come up for adjudication. I, personally, would predict that a case raising this question will probably reach the Supreme Court sometime during the next fifteen years or so, and that the court will probably decide that freedom of speech does, at least in general terms, embrace freedom of scientific inquiry. But that is pure guesswork at the moment, and in any event the Court might well write in a lot of small print limiting the application of the phrase "whatever scientific research one thinks fit."

Second, First amendment or no First amendment, the possession of a right is, as every first-year law student quickly learns, not the same as the exercise of that right. If we seek to exercise our constitutionally protected rights in a manner and a situation in which there is "clear and present danger" of public harm, the public authorities are perfectly entitled to intervene and place restraints on that exercise, as Justice Holmes reminded us. The First amendment does not authorize us to panic a crowded theater by shouting, "Fire!" where there is no fire. Given the degree of disagreement between scientists over the scale and nature of the risks involved in a DNA case, the Cambridge City Council quite reasonably apprehended the possibility of public harm, and it would have been negligent of them not to have intervened.

Why has there ever been any doubt about this? The reason is, I believe, because in one crucial respect the recombinant DNA is, in fact, a historic first. There have been previous cases in which the actual conduct of scientific experiments posed risks to the human beings directly involved. Hence all the current concern in recent years about the ethical review of research involving human subjects.

Again, there have been previous cases in which effects of applying the results of scientific research on a large scale subsequently posed a serious threat to public health or welfare. The whole controversy about the use and abuse of nuclear technology is an obvious example of this.

Plenty of secondary lines of scientific investigation, too, have required special constraints, for instance, those involving dangerous viruses or nerve gases or radioactive substances, and the scientists who do their research in these fields are by now perfectly accustomed to the fact that they have to place certain constraints and that indeed constraints are placed on them in their research by public authorities. But I can think of no prior case in which the actual conduct of fundamental experiments in a basic natural science itself directly posed a threat of general public harm. Even in the case of nuclear physics and the "artificial transmutation of the elements," as it was called when I was a boy, the direct effects of the initial experiments conducted by Rutherford and his colleagues at the Cavendish back in the early 1920s (not in 1942, as we were told last night), the direct effects of these experiments were entirely localized and involved no risk at all to the general public.

Why is the recombinant DNA case unique in this respect? In all previous situations involving the production by scientists of toxic substances or agents as part of a program of strictly basic research there was no difficulty in limiting the spread of those agents or substances. But the very heart of the DNA problem, it seems to me, is the suggestion that any "rogue" agents produced artificially in the course of research might have the power to multiply themselves and spread throughout the population at large, for example, by colonizing the human gut, so distinguishing themselves from, for instance, the minute quantities of artificially radioactive material produced by Rutherford's experiments and the like. It is this multiplication effect that is rightly perceived by outsiders as requiring special safeguards and as justifying a deliberate, conservative approach until it is clear that such safeguards are available.

Recombinant DNA Is a Permissible Field

To move now to the forbidden knowledge issue, it is evidently unwise for scientists to underestimate the power of the public imagination in this respect. The manner in which the recombinant DNA issue is being presented for public debate, both by journalists and by some scientists themselves, has been one that quite naturally awakened echoes of, for instance, the Faust legend and so provoked the kinds of public anxiety and suspicion that in earlier centuries confronted the alchemists–for instance, the suspicion that they were attempting to produce the *Homunculus*, that is, an artificial human embryo generated within their own alembic or retort out of lifeless raw materials. So it is necessary to correct some of the misunderstandings on which the current suspicion rests. Certainly I know of nothing in the way of DNA research that is actually in prospect as contrasted with speculative propaganda either pro or con which could even remotely answer to the *Homunculus* specification.

During the 1920s and 1930s, as I recall, the activities of the atom splitters at the Cavendish and elsewhere aroused something of the same frisson among the general public. Splitting the atom was suspected also of tampering impermissibly with nature's mysteries. Rutherford, in response, called his own little book of popular science "The Newer Alchemy." But it is clear by now that the whole field of atomic and nuclear physics covers a multitude both of good and of bad things. So the question is not whether to permit or outlaw all such work but rather what sorts of research we should do in this area, with what safeguards, and how shall we control the possible misapplication of the knowledge gained as a result.

I want to take the two parts of this issue in turn. First, any claim that all recombinant DNA research in itself involves impermissible tampering with the natural processes of organic evolution is surely far to general and undiscriminating. In itself the research does nothing of the sort. It could do so only if its by-products escaped into the biosphere or if the knowledge gained from the initial research were misapplied.

In any event, we have, as we have already been reminded, been intervening in the natural process of organic evolution for millennia. All culture, especially the domestication of plants and animals and their selective breeding, involves tampering with evolution. So, once again the question is not whether we are to begin doing it for the first time but rather how we are to be sure in this particular case that we are doing it for good rather than for ill.

Second, how are we to be sure of that? Again, I believe we have perfectly good models available that answer that question in principle, if not in detail. The principles may remain the same, but the details are always new. The issue was settled in principle, as I see it, in prehistory, and the outcome is enshrined in the legend of Prometheus. In its own time and in its own terms, discovering ways of producing fire artificially was as daunting as discovering techniques for producing artificial radioactivity, nuclear power, or synthetic forms of DNA today.

As people very soon came to recognize, the proper response

was not to outlaw the very use of fire. Rather it was to invent the legal concept of arson. This is known as replying to a four-letter word with a five-letter word. Rather, I say it was to invent the legal concept of arson and to develop effective legal sanctions, institutional mechanisms, public sentiments, and other practical safeguards against the misuse of fire.

It is very easy to express Arcadian sentiments, but I think it would be worthwhile, especially after a January in Chicago speculating about what it would be like to live without the artificial handling of fire. So, for all I can see, there is nothing forbidden or impermissible about the knowledge we could get from recombinant DNA research, provided that we are also taking seriously the practical question of the safeguards against foreseeable harm and even against unforeseeable harm and the eventual possible misuse of that knowledge.

47

Cloning Serves the Interests of Those in Power

Jonathan Beckwith

I have been doing research in bacterial genetics for the last twelve years at Harvard Medical School, and I am a member of Science for the People. Over the last couple of years, we have been discussing in our laboratory how recombinant DNA technique could make certain of our experiments much easier to do. However, as a result of these discussions we decided not to use this technique at all. This is not because the particular experiments we were talking about could be thought of as health hazards in any way. Rather, my reasons were that I do not wish to contribute to the development of a technology which I believe will have profound and harmful effects on this society. I want to explain why some of us have arrived at this decision.

In 1969, a group of us in the laboratory developed a method for purifying a bacterial gene. We took that opportunity to issue a public warning that we saw developments in molecular genetics were leading to the possibility of human genetic engineering. While we saw genetics progressing in this direction, we had no idea how quickly scientists would proceed to overcome some of the major obstacles to manipulating human genes. The reports on the use of recombinant DNA technology, beginning in 1973, represented a major leap forward. The result is that geneticists are now in the position to purify human genes. And proposals have already been put forward for the setting up of "mammalian DNA banks." Further, techniques are being developed that will allow reintroduction of those genes into mammalian cells. These steps appear perfectly feasible.

There are still some barriers left to introducing genes into

This appeared as "Present and Future Abuses" in *Research with Recombinant DNA,* National Academy Sciences, Washington D.C., 1977.

human cells, organs, or embryos at the proper time or in the proper way. But these goals are not inconceivable and they may be achieved very rapidly. Whatever the current state of knowledge, to claim that the possibilities of genetic engineering of humans with this technique is far off is to totally ignore the history of this field.

In 1969, most scientists pointed to the impossibility of purifying human genes and claimed that such developments were at least decades off. In fact, they were four years off. Let's not be fooled again. Just as suddenly as recombinant DNA appeared on the scene, breakthroughs in "genetic surgery" may appear.

And when the day arrives in the near future when geneticists have constructed a "safe" vector for carrying mammalian genes into human cells, others will begin to use it for human genetic engineering purposes. There has already been at least one reported case in which there were direct attempts to cure a genetic disease in human beings with virus-carried genes, and in human cells.

But why be concerned about human genetic engineering? There are certainly many individuals and groups that have ethical or religious objections to any intervention of this kind in human beings. Possibly after widespread discussion within a society those objections might predominate. I, personally, do not necessarily view all human genetic intervention as inherently to be opposed. But, I would rather point today to some concrete dangers of the development of recombinant DNA research by examining the scientific, social, and political context in which it is proceeding. For that reason, much of what follows will speak to those issues rather than directly to recombinant DNA.

Genetic Engineering Has Begun

In the last ten or fifteen years, there have been major advances in a number of areas of genetics which bring us to a situation today, in which genetic engineering already is underway. These include a variety of types of genetic screening programs in which it is possible to identify genetic differences between people by examining cells of individuals. The approaches are:

1. Amniocentesis—where the cells of a fetus obtained from a pregnant woman can be examined for genetic variations. In a small number of cases, these variations are known to cause serious health problems, and suffering may be eliminated by giving the parents the option of aborting such a fetuses.

2. Postnatal screening—when infants are screened after birth for genetic differences. Again, in a small number of cases, those variations may cause disease and treatment may be provided.
3. Adult screening—where prospective parents can be advised of the likelihood of their bearing children who might carry particular genetic variations.

While each of these programs have proved beneficial to some individuals, they have also encountered problems, been controversial, and, in some cases, caused suffering to those screened. In addition, all of these programs raise the basic question of who is deciding who is defective, or even, who shall live?

There are other developments which have received much attention in the press, e.g., the possibility of cloning genetically identical individuals and the attempts to grow fertilized eggs in the test tube and then implant them in a woman's uterus.

At the same time these developments in genetic technology were taking place, there were also a growth in studies in human behavioral genetics. In the last ten years, there as been a resurgence of supposedly "scientific" research that claims to explain many of our social problems as being due to genetic differences between people. For instance, there are attempts to say that the inequality which exists in this country or the lower achievement of various groups, particularly blacks, is due to inferior genes. Or the proposals that criminality might be explained by genetic differences between the criminal and the noncriminal—the case of the XYY male. (By the way, on of the reasons that I suggest genetic engineering is already under way, is that XYY fetuses have been aborted after detection by amniocentesis.) In both these cases, the scientific evidence has been shown to be nonexistent and, in some cases, fraudulent. In addition, there are the more recent attempts in the field of sociobiology to claim biological and genetic evidence to justify the lower status position of women in this society. It is a disgrace that this government continues to support such shoddy, groundless, and ultimately harmful research.

Genetic Arguments Are Used Against the Workers

These genetic theories and the problems wit genetic screening programs did not arise in a social and political vacuum.

They have followed a period of intense social agitation and social disruption in the United States. After black, other minority groups, the poor, and women demanded a greater share of the wealth and power in this society, the response arose that such equality is genetically impossible. The ghetto uprisings and other violent confrontations that occurred during this period are explained as being due to people whose genes are "off." The demands of the women's movement are met with the answer that women are genetically programmed for the roles they now occupy.

Another more recent example of this genetic approach to social problems lies in the field of industrial susceptibility screening. Arguments have been appearing in the scientific literature and elsewhere that occupational diseases, caused by pollutants in the workplace, can be ascribed not to the pollutants themselves, but to the fact that some individuals are genetically more susceptible to the pollutants than other individuals. So the argument goes, the solution is not getting rid of the pollutants, but rather, for example, simply not hiring those individuals who are thought to carry the genetic susceptibility. Now, clearly, whenever it is possible to warn someone of dangers he or she may face, that information is important. However, what is blatantly ignored by those promoting this area of research is that, in almost every case, nearly everyone in the workplace is at some degree of increased risk because of exposure, for instance, to asbestos fibers. Yet, already, there are headlines in the newspapers such as the following: "Next Job Application May Include Your Genotype." A Dow Chemical plant in Texas has instituted a large-scale genetic screening program of its workers. Rather than cleaning up the lead oxide in General Motors plants, women of child-bearing age are required to be sterilized if they wish employment. It is a genetic cop-out to allow industries to blame disease on the genetically different individual rather than on their massive pollution of the workplace and the atmosphere. This is the epitome of "blaming the victim."

The end result of these genetic excuses for society's problems is to allow those in power in the society to argue that social, economic, and environmental changes are not needed—that a simpler solution is to keep an eye on people's genes. And thus the priorities are determined. For example, major funding goes to genetics research and into viral causes of cancer, and a pittance to occupational health and safety. This distorted perspective is reinforced by the emphasis and the publicity that recombinant DNA research has achieved with its claim for solving problems whose solution are mainly not in the realm of genetics. Typical of the claims made by those promoting this area is a statement by my fellow panelist David Baltimore: "How much do we need recombinant DNA? Fine, we can do without it. We have lived with famine, virus and cancer, and we can continue to."

That is not a neutral or apolitical statement. The sources of famine and disease lie much more in social and economic arrangements than in lack of technological progress. Aside from the incredible claims for the benefits of recombinant DNA, this statement essentially opts for the status quo. Social problems, such as famine and disease, are taken out of the arena of political action and sanitized behind the white coat of the scientist and the doctor. Of course, we might have both social and medical approaches to such problems going on at the same time. But given the current struggle over solutions to these problems, such statements can only provide weapons to those who would like to maintain present power relationships and profits. What is opted for are the technological fixes, in this case, the genetic fix.

Recombinant DNA–The Genetic Fix

Let me give you some examples of how we may move from the present technological fix to the genetic fix, once recombinant DNA techniques have provided the tools. In the United States over the last few years, approximately 1 million school children per year have been given drugs, usually amphetamines, by the school systems, in order to curb what is deemed disruptive behavior in the classroom. It is claimed that these children are all suffering from a medical syndrome, minimal brain dysfunction, which has no basis in fact—no organic correlate. Now, clearly there are some cases of children with organic problems where this treatment may well be important. But in the overwhelming majority of cases the problems are a reflection of the current state of our crowded schools, overburdened teachers and families, and other social problems, rather than something wrong with the kids. Imagine, as biochemical psychiatry is providing more and more information on the biochemical basis of mental states, the construction of a gene that will help to produce a substance in human cells which will change the mental state of individuals. Then, instead of feeding the kids a drug every day, we just do some genetic surgery and it's over.

Don't forget that introducing genes into humans—genetic engineering—results in permanent changes. There is no way to cut the genes out. It's irreversible. At least, when protests mounted in certain schools against the drugging of kids, the treatment could be stopped. That's not the case with the genetic solution. There's no going back.

Another example: A current idea, again without scientific foundation, is that aggression is determined by hormone imbalance. Males, it is said, are more aggressive than females because of the hormone testosterone or the absence of presumed female hormones. As a result, patients in mental institutions deemed aggressive are treated with the presumed female hormones. But recently it has been discovered that there are genes in bacteria that will break down testosterone. Wouldn't it be a simpler, less costly approach to introduce such genes (in a functional state) into the "aggressive" patient. Maybe even social protest can be prevented that way. But what are the sources of aggression in this society? Isn't it possible that rather than hormone imbalance, it is social and economic imbalance—unemployment, racism, and do forth—that spurs many people to "aggressive" behavior? And, while we're on the subject, would such genetic surgery be used on those in leadership positions in the society responsible for such atrocities as the Indochina war?

Similar approaches could be used to argue for gene therapy on fetuses, infants, or on the workers themselves so that they can work in factories with high vinyl chloride levels. Given the sophistication of the new technologies, a new eugenics era may do even greater damage than the earlier eugenics movement (1900–1930).

Genetic Engineering Threatens the Powerless

I would like to add a component to the benefit-risk discussion of recombinant DNA that has, for the most part, been ignored. This component is the risk of human genetic engineering to those without power in this society. Given the present social context I believe these consequences are inevitable. It is not just the particular evils and damage to individuals I have mentioned in my scenarios that concern me. The dramatic developments in this field, and the publicity they have received and will continue to receive, is already reinforcing the focus on the genetic fix. On the one hand, an atmosphere is being generated in which a variety of genetic approaches to social problems is accepted. And, as a corollary, social, political, and economic changes are deemphasized. The priorities of the society cannot be allowed to be dictated by the technocrats and their technology. On the contrary, technologies must be developed only after social decisions that they are wanted and needed.

On this basis, I believe we should seriously whether recombinant DNA research should be pursued at all.

XI

Can Cloning Be Banned?

Introduction to Part XI

Ever since Dolly, the main issue of practical policy arising from the possibility of human cloning is whether it should be banned outright. Although political sentiment was at first strongly in favor of such a ban, there may be insurmountable practical difficulties in the way of a prohibition which is effective, which is constitutional, and which does not cause enormous injury to extremely valuable research work.

Charles Krauthammer (Reading 48) contends that any attempt to prevent the development of cloning technology is doomed to failure. Eliot Marshall describes in Reading 49 how biomedical groups have reacted to the possibilities of legislative prohibitions. Vincent Kiernan (Reading 50) reports on the rejection by the Senate of one attempt at criminalization of cloning. An account of the European legislation designed to outlaw human cloning is given by Teresa Malcolm (Reading 51). Reading 51 reproduces part of Senator Ted Kennedy's Senate speech opposing one piece of legislation he judged to be too restrictive. Susan Wolf (Reading 53) presents arguments that a cloning ban would be unconstitutional, and calls for more flexible regulation. Finally, in Reading 54, Andrea Bonnicksen describes the political "rush" to ban a type of human cloning which was, she maintains, not at all imminent, and advocates a more nuanced regulatory approach.

48

Nothing Can Stop the Cloning of Humans

Charles Krauthammer

One doesn't expect Dr. Frankenstein to show up in wool sweater, baggy parka, soft British accent and the face of a bank clerk. But there in all banal benignity he was: Dr. Ian Wilmut, the first man to create fully formed life from adult body parts since Mary Shelley's mad scientist.

The creator wore chinos. Wilmut may not look the part, but he plays it. He took a cell nucleus from a six-year-old ewe, fashioned from it a perfect twin—adding the nice Frankenstein touch of passing an electric charge through the composite cell to get it growing—and called it Dolly.

Dolly, the clone, is an epochal—a cataclysmic—creature. Not because of the technology that produced it. Transferring nuclei has been done a hundred times. But because of the science. Dolly is living proof that an adult cell can revert to embryonic stage and produce a full new being. This was not supposed to happen.

It doesn't even happen in amphibians, those wondrously regenerative little creatures, some of which can regrow a cut-off limb or tail. Try to grow an organism from a frog cell, and what do you get? You get, to quote biologist Colin Stewart, "embryos rather ignominiously dying (croaking!) around the tadpole stage."

And what hath Wilmut wrought? A fully formed, perfectly healthy mammal—a mammal!—born from a single adult cell. Not since God took Adam's rib and fashioned a helpmate for him has anything so fantastic occurred.

What, then, was the reaction to this breakthrough of biblical proportions?

There is a mischievous story (told mostly in England) that a

This appeared as "A Special Report On Cloning" in *Time*. v149 (Mar. 10 1997), 1997. pp. 60–7.

leading Scottish newspaper reported the Titanic sinking with the headline GLASGOW MAN LOST AT SEA. Well, here was a story that deserved the headline MAN CREATES LIFE. And how does it play? A *Wall Street Journal* headline urgently asks, WHO WILL CASH IN ON BREAKTHROUGH IN CLONING? (Answer: "Tiny company could emerge a big winner.") The President of the U.S. calls for a committee of experts to gather and pull their beards.

And the *New York Times*, in a lovely coda to its editorial titled CLONING FOR GOOD OR EVIL, advises that "society will need to sort through what is acceptable and what is the nightmare beyond."

Well, yes. The most portentous scientific achievement since Alamogordo will need a weighing of pros and cons. No kidding.

And, no doubt, the pro-and-con weighing, the pontificating and the chin pulling will now go into high gear. Wilmut will spawn more ethics conclaves than cloned sheep. No matter. There is nothing to stop cloning, not even of humans.

What the politicians do not understand is that Wilmut discovered not so much a technical trick as a new law of nature. We now know that an adult mammalian cell can fire up all the dormant genetic instructions that shut down as it divides and specializes and ages, and thus can become a source of new life.

You can outlaw technique; you cannot repeal biology. And even the outlawing of this technique—Britain, for example, forbids the cloning of humans—will fail. It is too simple, too replicable. No amount of regulation by the FDA or the NIH or even the FBI will stop it.

Why? Not just because it is so easy, but because its potential for good is so immense. The study of cloning can give the world deep insights into such puzzles as spinal cords, heart muscle and brain tissue that won't regenerate after injury, or cancer cells that revert to embryonic stage and multiply uncontrollably. Replicating Wilmut's work will elucidate what he along the way did right that nature, in these pathologies, does wrong.

Of course, the potential for evil is infinitely greater. But there will be no stopping that either. Ban human cloning in America, as in England, and it will develop on some island of Dr. Moreau. The possibilities are as endless as they are ghastly: human hybrids, clone armies, slave hatcheries, "delta" and "epsilon" sub-beings out of Aldous Huxley's *Brave New World*.

But you don't have to be mad to be tantalized. Being human will do. Think of it: what Dolly—fat, insensible Dolly—promises is not quite a second chance at life (you don't reproduce yourself; you just reproduce a twin) but another soul's chance at your life. Every parent tries to endow his child with the wisdom of his own hard-earned experience. Here is the opportunity to pour all the accumulated learning of your life back into a new you, to raise your exact biological double, to guide your very flesh through a second existence.

Oh, the temptation to know what might have been. Or to produce an Einstein, a Dr. King, for every generation. Or to raise a Jefferson in a clearing, a cross between Jurassic Park and Williamsburg, an artificial environment re-creating 18th century Virginia. Create, nurture and wait. Then bring him out one day, fully grown, to answer the question of the ages: What would Jefferson do today?

49

Biomedical Groups Mobilize Against Prohibition of Cloning

Eliot Marshall

The research community is not known as a powerful political force. But last week, biomedical groups demonstrated surprising muscle—and lobbying tactics that would have made the most seasoned Washington insider proud—when they derailed legislation moving on a fast track through the U.S. Senate. Their target: a bill that would have made it a crime to clone humans with the technology used last year to make Dolly, the world's most famous sheep.

The research groups took on quite a challenge. The notion of outlawing human cloning has widespread popular appeal, and the bill itself had the backing of the most powerful man in the Senate, Majority Leader Trent Lott (Republican–Mississippi). Indeed, its supporters were so sure they had a winner that they tried to bring the bill straight to the Senate floor, bypassing committee hearings and debate. But Lott's tactic, it turned out, was a mistake. Opponents—including scientific societies, industry organizations, patient advocacy groups, and 27 Nobel Prize winners—argued that the bill would block basic biomedical research as well as human cloning. They won over enough senators, including such unlikely bedfellows as Strom Thurmond (Republican–South Carolina) and Edward Kennedy (Democrat–Massachusetts), to put off a vote.

It was a sharp loss for the Republican leadership, forcing Lott to withdraw the legislation. Says one Republican aide: "Nobody takes pleasure in handing the Majority Leader a defeat." The battle is far from over, however, as the Senate bill may reemerge and other anticloning bills are also circulating in

---

This appeared as "Biomedical Groups Derail Fast-track Anticloning Bill" in *Science*, v279 n5354, February 20, 1998, pp.1123–1124.

Congress. But last week's showdown set the stage for a coming debate—and provided an object lesson in the hardball politics of biomedical policy.

Momentum for a cloning ban began building in January, as members of Congress returned from a long winter break during which they heard that Chicago physicist Richard Seed was trying to raise money to clone humans (*Science*, 16 January, p. 315). President Clinton added to the clamor in his State of the Union Address on 27 January by calling for legislation to block human cloning. Some members of Congress wanted to seize the initiative. According to staffers of several biomedical interest groups, conservative Senators Judd Gregg (Republican–New Hampshire) and Kit Bond (Republican–Missouri) met with Lott in late January to push for speedy action. Bond suggested they support a bill he had introduced last year during the Dolly furor. It would outlaw not just the creation of humans by cloning, but any research involving human eggs and the process of somatic cell nuclear transfer. This process—which was used to create Dolly—removes the nucleus of an egg and replaces it with the nucleus of another cell. The embryo is then stimulated to grow.

The conservatives gained an important ally in Senator Bill Frist (Republican–Tennessee), a former transplant surgeon who heads the public health and safety subcommittee. With Lott's blessing, Frist and Bond co-sponsored a new version of Bond's bill (S. 1601), proposing to make human somatic cell nuclear transfer illegal, punishable by a ten-year prison sentence. The bill would also create a 25-member national commission to report on ethical issues in biomedicine. Lott himself introduced the bill on 3 February.

Biomedical organizations quickly mobilized. Many researchers had expressed concern about Bond's bill last year, arguing that the cloning procedure it would outlaw might be used to produce not just embryos but primordial stem cells, which future technology might convert into transplantable bone marrow, skin, or other tissue. They also worried that the bill would broaden and extend the current moratorium on research done with human embryo cells. Sean Tipton, a spokesperson for the American Society for Reproductive Medicine, broadcast an appeal for help over the Internet. Tipton warned that under the guise of preventing human cloning, some members of Congress were making "a serious attempt … to permanently enact an embryo research ban." He asked the community to "stand up and be counted" in the name of "the freedom of scientific inquiry."

The Pharmaceutical Research and Manufacturers of America (PhRMA), an association representing large drug companies, joined the fray. It sponsored a press conference on 6 February at which PhRMA scientist Gillian Woollett warned that passage of S. 1601 could cast "a pall over a whole area of research," scaring away researchers who might use somatic cell nuclear transfer to develop skin cells for burn victims, bone marrow for cancer patients, and neuronal cells for people with spinal cord injuries. Herbert Pardes, dean of the Columbia University College of Physicians and Surgeons, and Heather Fraser, a patient spokesperson for the Cystic Fibrosis Foundation, argued that outlawing research would set a dangerous precedent. In parallel, the Biotech Industry Organization and about 70 patient advocacy groups and professional societies added their voices to the chorus.

But perhaps the weightiest blow was delivered on 9 February, when the American Society for Cell Biology distributed a letter signed by 27 Nobelists, including several from outside biology, such as economist Kenneth Arrow and physicist Douglas Osheroff of Stanford University. It declared "a broad consensus" in favor of banning human cloning through a voluntary moratorium. If anticloning legislation must be passed, they said, it should apply only to the creation of human beings, not embryos, and should "not include language that impedes critical ongoing and potential new research." Speaking for the group, biochemist Paul Berg of Stanford told The *New York Times*, "The Bond-Frist bill is clearly going to block very important research."

The opponents already had two important allies in the Senate: Senators Kennedy and Diane Feinstein (Democrat–California). These two had introduced an alternative bill (S. 1602), crafted with the advice of biomedical groups. The Feinstein-Kennedy bill would, for ten years, make it illegal to implant into a woman's uterus an embryo created by cloning techniques such as somatic cell nuclear transfer. But their bill would not outlaw research on human somatic cell nuclear transfer. When Lott tried to bring the Bond-Frist bill to a vote, Feinstein and Kennedy began a filibuster.

A filibuster can be broken only if at least 60 senators vote to end debate. Lott appealed to his fellow Republicans to back him on such a vote—a plea that would normally get automatic support.

This time, it didn't. Senator Connie Mack (Republican–Florida), a cancer survivor and champion of biomedical research, convinced that it would be a mistake for the Senate to vote on the bill without hearings, persuaded 11 other Republicans to join him in blocking Lott. They added their number to 42 Democrats. The majority included Thurmond, who spoke with emotion of his hope that basic research might help his diabetic daughter, urging that no laws be placed in the way. Lott's motion failed by a vote of 54 to 42, and the bill was put aside. Says Tipton: "We've dodged the first bullet."

The next day, in the House of Representatives, the Commerce Committee began reviewing a range of proposals for a ban on cloning. Neither the Republican nor Democratic members seemed in a hurry to send legislation to the floor, however, as they heard from religious leaders and probed the meaning of words such as "embryo," "somatic cell," and "human life." The Senate Republican leadership had not made a decision at press time whether to send the Bond-Frist bill to committee for additional review. A spokesperson in Bond's office said only that the bill had been withdrawn from debate, but could be brought back "at any time." The legislation has been shunted off the fast track, perhaps, but not derailed.

50

The Senate Rejects a Bill to Ban Human Cloning

Vincent Kiernan

The Senate last week rejected legislation that would have made it a felony to clone a human being. Scientists had complained that the legislation would have interfered with research.

Republican lawmakers in the House of Representatives, however, are preparing to try to pass similar legislation, perhaps as early as this month.

The bill that was defeated in the Senate would prohibit the use of somatic-cell nuclear-transfer technology with human cells. Scottish researchers employed that technique to clone an adult sheep last year.

"We were able to mobilize a tremendous amount of scientific expression very quickly," said David Korn, senior vice-president for biomedical and health-sciences research at the Association of American Medical Colleges, one of several groups that lobbied against the cloning ban. "I'm very gratified, but I realize it's just the first skirmish in a long war."

In December, Richard Seed, a physicist, set off a furor by announcing plans to use the technology to try to clone humans. When Congress reconvened in January, Republican leaders promised quick action to ban human cloning.

But when they brought to the floor a bill sponsored by Senator Christopher S. Bond, a Missouri Republican, opponents used a parliamentary maneuver to prevent debate.

Last week, supporters of the bill sought to force a debate on it but failed to muster the three-fifths majority required to override the obobjections. Forty-two Senators supported

---

This appeared as "Senate Rejects Bill To Ban Human Cloning" in *Chronicle of Higher Education*, v44 n24, pp.A40–A41, Feb 20, 1998.

consideration of the bill, and 54 opposed it.

"Those who support human cloning may have won this round, but the battle has just begun," Mr. Bond said after the vote.

Indeed, the day after the Senate vote, the House Commerce Subcommittee on Health and Environment held a lengthy hearing on banning human cloning. Last August, the House Science Committee approved a bill that would prohibit the use of federal funds to produce human clones. The measure is pending in the House.

At the House hearing last week, republican lawmakers said they opposed cloning for research purposes because cloned human embryos would be destroyed in the process. "The issue of human cloning is fundamentally and undeniably about life," said Thomas Bliley, the Virginia Republican who chairs the House Commerce Committee. Democrats said research using cloning should be allowed to proceed. Sherrod Brown of Ohio, the senior Democrat on the Health subcommittee, said he would introduce a bill to impose a moratorium on producing humans through cloning.

Under Mr. Bond's bill, also known as the Human Cloning Prohibition Act, someone found guilty of using the cloning technology with human cells could face fines and 10 years in prison.

The chief foes of that proposal—Democratic Senators Dianne Feinstein of California and Edward M. Kennedy of Massachusetts—have proposed their own legislation, which would ban the placement of cloned human embryos in a uterus, but would allow the clones' use in research. Their bill includes fines but no prison terms for violations and would lapse in 10 years; Mr. Bond's bill has no sunset clause.

Although Democrats led the opposition to Senator Bond's bill, some Republicans—including several conservatives—also opposed it. Among them were Senator Strom Thurmond of South Carolina. He told his colleagues in last week's debate that his daughter suffers from diabetes. Research using cloning, he said, might provide new insights into diabetes as well as Alzheimer's disease, sickle-cell anemia, multiple sclerosis, and muscular dystrophy.

"I am concerned that this bill may be so broadly written that it may interfere with future promising research," Mr. Thurmond said. He said the Senate should hold hearings on cloning legislation before enacting a bill.

Senator Bond and others, however, played down the bill's potential effect on science. "Let me be clear: This bill does not stop existing scientific research," said Mr. Bond.

Scientists Are Challenged on Their Claims about Harm to Research

Senator William H. Frist, a Tennessee Republican who is a heartand-lung-transplant surgeon, challenged scientists to cite a single "peer-reviewed study" that would have been impossible if the bill had been in effect. "No longer can we divorce science from ethical considerations," Dr. Frist said.

In the days before the vote on Mr. Bond's bill, scientific associations had mounted a major lobbying effort against it, and the Clinton Administration had announced its opposition. Besides the Association of American Medical Colleges, groups opposing the bill included the American Cancer Society, the American Council for Cell Biology, the Biotechnology Industry Association, and the Federation of American Societies for Experimental Biology. Twenty-seven Nobel Prize-winning scientists also wrote to President Clinton and members of Congress to oppose the bill.

Researchers from the National Institutes of Health produced a paper that described various possible uses of the technology in both basic research and medical therapies. "Somatic-cell nuclear transfer holds many diverse and important possibilities to significantly prevent, treat and maybe cure disease. All of these possibilities can be accomplished without using this technology to create a human being," concluded the paper, which was circulated by opponents of Mr. Bond's bill. The paper does not bear the names of its authors at N.I.H., because President Clinton has endorsed a ban on the technology, and they did not want to be publicly identified as opposing him, one source said.

The prevailing attitude in the scientific community appeared to be that federal regulations governing cloning would be preferable to legislation, but that if legislation was inevitable, the Kennedy–Feinstein bill was preferable to Mr. Bond's.

Senator Bond's bill is "beyond redemption," said Dr. Korn, of the medical-college association. "It's poorly written, illogical, and just bad."

By contrast, Dr. Korn said, the Feinstein–Kennedy bill is "less noxious but still suffers from the ambiguities of the English language." For example, he said, it would bar cloning of "mature" cells but also could be construed to forbid certain procedures in reproductive medicine that manipulate human embryos to eliminate cells that carry genetic diseases.

"Legislation is a very poor way of dealing with issues of scientific and medical progress," said David Baltimore, a Nobel Prize-winning virologist and president of the California Institute of Technology. Once in place, a law is difficult to change, he argued. By contrast, a federal agency can more easily revise regulations to keep them up to date with the fast pace of developments in science, he said. "What looks to be revolutionary one day is common practice the next."

Dr. Baltimore said human cloning may someday be deemed morally acceptable, as the technology comes to be seen as an aid to infertile couples. He said the government should follow the example it set in the 1980s in its regulation of genetic engineering. Although at the time there were calls to outlaw the research, the government instead issued regulations that restricted it. Over the years, as fear of the technology ebbed, the restrictions were eased. "It is now an accepted technology of medical science," he said.

Gradually relaxing those regulations helped encourage the development of a robust biotechnology industry in the United States, Dr. Baltimore added. On the other hand, European nations banned the technology outright, and consequently the European biotechnology industry consistently has lagged behind the United States, he said.

"The idea of making this a federal crime really seems to be bizarre," said Gregory E. Pence, a professor of philosophy at the University of Alabama at Birmingham and author of a new book, Who's Afraid of Human Cloning? (Roman and Littlefield). "Everything at this point is based on fear and ignorance."

Some Scientists Support a Cloning Ban

Many of the fears about cloning—that a scientist would be "playing God," or that a clone would have no identity apart from its "parent"—are based on a misguided view of "genetic reductionism," in which persons are totally determined by their genetic makeup, he said.

But some scientists support a ban. Marie DiBerardino, an emerita professor of biochemistry at Allegheny University of the

Health Sciences, in Philadelphia, said a ban was justified because of the risk that a researcher might defy a voluntary moratorium.

"There are these characters around who are going to go ahead" with human cloning despite the strong public sentiment against it, she said. "We have to permit legislation."

A ban on human cloning would not hamper basic research, she argued. "That's a red flag that very liberal scientists are holding up." Research into cloning could proceed with other animals, such as mice and cows, she added—scientists "don't have to be playing with human material."

Robert G. McKinnell, a professor of genetics and cell biology at the University of Minnesota who clones frogs as part of his research into cancer, said he would support legislation that banned the production of human beings through cloning. The world already has enough people, he said.

"We have real problems, like heart disease and diabetes," he added. "Making more people at the laboratory bench does not in any way solve any problems."

He said, however, that he would oppose legislation against the cloning of human cells for therapeutic purposes—ones that did not lead to the birth of humans—such as using the technique to produce bone marrow for use in cancer therapies. Such uses of cloning are little more than science fiction now, he said, but may become feasible in the future.

Others said a ban on cloning would be premature, because the technology itself is unproven. In the January 30 issue of Science, Vittorio Sgaramella, of the University of Calabria, in Italy, and Norton D. Zinder, of Rockefeller University, expressed skepticism about the Scottish sheep experiment, noting that the feat had yet to be repeated.

In a letter to the journal, they noted that the Scottish researchers had required more than 400 attempts to produce a single cloned sheep. That amounts to "an anecdote, not a result," the two scientists wrote.

51

Europe Bans Human Cloning

Teresa Malcolm

Officials from 19 European nations signed an agreement January 12 banning human cloning.

"At a time when occasional voices are being raised to assert the acceptability of human cloning and even to put it more rapidly into practice, it is important for Europe to solemnly declare its determination to defend human dignity against the abuse of scientific techniques," said Daniel Tarchys, secretary general of the Council of Europe, which drew up the agreement.

Church Groups Are Dismayed by Seed's Proposals

The ban was prompted in part by last week's announcement by independent scientist Richard Seed that he was ready to set up a clinic to clone human babies.

Church groups and other religious experts on bioethics have expressed dismay at the Chicago-area physicist's recent announcement that he plans to begin work on human cloning.

Seed has said "cloning ... is the first serious step in becoming one with God." The Rev. Ronald Cole–Turner of Pittsburgh Theological Seminary and author of Human Cloning: Religious Responses, said such statements by Seed are "outrageous."

"Almost without exception religious leaders say they have problems with cloning," Cole–Turner said. "Some say 'never' to human cloning. Others, like myself, wouldn't say 'never,' but are very apprehensive."

Though worried about the "theological" repercussions of using cloning "to make babies," the Rev. Ted Peters, a Lutheran minister and a research associate with the Berkeley, California-based Center for Theology and the Natural Sciences, said "some benefit" may come from using human cloning in medical

---

"Europe Bans Human Cloning" in *National Catholic Reporter*, v34 n13, pp.7, Jan 30, 1998

procedures such as organ transplants. He cautioned that such research should proceed very carefully—even with cloning that wouldn't produce any children.

52

Vital Medical Research Is Threatened by a Proposed Cloning Ban

Edward Kennedy

[The legislation before the Senate] isn't a bill to ban a brave new world of mass production of cloned human beings. It is not legislation to stop wealthy individuals from reproducing themselves at will in an unscrupulous and unethical attempt to achieve a kind of immortality. Instead, this legislation bans the actual technology used in human cloning research—the technology that could be used to create cures for cancer, diabetes, spinal cord injuries, arthritis-damaged joints, birth defects, and a host of terrible neurological diseases like Alzheimer's disease, Parkinson's disease, Lou Gehrig's Disease, and multiple sclerosis.

Every scientist in America understands the threat this legislation poses to critical medical research. Every American should understand it, too. ... Congress can and should act to ban cloning of human beings during this session. But it should not act in haste, and it should not pass legislation that goes far beyond what the American people want or what the scientific and medical community understands is necessary and appropriate.

... I have no doubt that responsible legislation to ban the production of human beings by cloning can come through committee and markup and be passed into law during this session of Congress. ...

... [The American Association of Medical Colleges compare this legislation] to ban not just cloning of human beings but use of the technique itself to the ill-considered attempts to ban recombinant DNA techniques in the '70s.

They state,

Like the recombinant DNA debate, the scientific techniques involved in cloning research hold great promise for our ability to treat and manage myriad diseases and disorders—from cancer and

---

This appeared in the *Congressional Record* on February 9, 1998 p. S513–14.

heart disease, to Parkinson's and Alzheimer's, to infertility and HIV/AIDS.

As of this morning, the letter had been signed by 71 distinguished organizations, from the American Academy of Allergy, Asthma, and Immunology, to the Association of American Cancer Institutes to the Parkinson's Action Network—and the list continues to grow ...

Is this really what the Senate or the American people want, Mr. President? To lose ground in the battle against deadly and disabling human diseases? I don't believe so.

More than 120 scientific and medical organizations have expressed opposition to the Lott-Bond bill or concerns about prohibition on legitimate cloning research as the result of ill-conceived or over-broad legislation ...

This is one of the most important scientific and ethical issues of the 21st century.

The Lott-Bond bill does not just ban cloning of human beings, it bans vital medical research related to cloning—research which has the potential to find new cures for cancer, diabetes, birth defects and genetic diseases of all kinds, blindness, Parkinson's disease, Alzheimer's disease, paralysis due to spinal cord injury, arthritis, liver disease, life-threatening burns, and many other illnesses and injuries. ...

It does not just ban the technology for use in human cloning. It bans it for any purpose at all.

That means scientists can't use the technology to try to grow cells to aid men and women dying of leukemia. They can't use it to grow new eye tissue to help those going blind from certain types of cell degeneration. They can't use it to grow new pancreas cells to cure diabetes. They can't use it to regenerate brain tissue to help those with Parkinson's disease or Alzheimer's disease. They can't use it to regrow spinal cord tissue to cure those who have been paralyzed in accidents or by war wounds.

Congress should ban the production of human beings by cloning. But we should not slam on the brakes and stop scientific research that has so much potential to bring help and hope to millions of citizens.

53

A Proposed Cloning Ban Unconstitutionally Restricts Scientific Research

Susan M. Wolf

In its report on cloning, NBAC recommended a ban of unprecedented scope.[1] Based on commission consensus that human cloning would currently be unsafe, NBAC called for congressional prohibition throughout the public and private sectors of all somatic cell nuclear transfer with the intent of creating a child. President Clinton promptly responded by proposing legislation to enact such a ban for five years.

NBAC was wrong to urge a ban. Cloning undoubtedly warrants regulation. But the ban proposed will not yield the sort of regulation required. Instead, it will reduce cloning to a political football in Congress, raise serious constitutional problems, and chill important research. NBAC defends its ban as a limited one, prohibiting somatic cell nuclear transfer (not all forms of cloning), when used to create a child (not in research), and for three to five years (not indefinitely). A congressional ban, however, is likely to be far broader.

NBAC erred by taking cloning out of context. Like any technology, cloning needs to be safe before used. But that counsels regulation, not a ban, which merely slows development of safe procedures. And cloning demands we deal with issues beyond safety on which NBAC achieved no consensus, issues bound up in the ethics of human experimentation and reproductive technologies.

A better approach would extend human subjects protection into the private sphere and regulate reproductive technologies effectively, with a central advisory body for novel issues such as cloning. By failing to tackle private research and reproductive

This appeared as "Ban Cloning? Why NBAC is Wrong," *The Hastings Center Report*. v27 (Sept./Oct. 1997), 1997. pp. 12–15.

technologies, NBAC avoided the real job and instead proposed an isolated and misguided response to cloning.

The Regulatory Challenge

Human cloning clearly requires regulation. Indeed, some regulation already applies. President Clinton has barred all federal funds for cloning, covering both research and clinical application.[2] Earlier prohibitions on the use of federal money to create human embryos for research purposes would also impede cloning research with federal funds.[3] And federal regulations protecting human subjects would seem to block cloning in research covered by those regulations because cloning remains unsafe, at least for now.[4] This leaves two regulatory gaps that properly troubled NBAC: private sector research outside federal oversight and private clinical activity, especially infertility programs using reproductive technologies.

But by responding to these worries with a congressional ban, NBAC missed the target. Protecting human subjects in private research and regulating reproductive technologies are both long overdue. A ban on cloning just suppresses one technology, while these two systemic problems guarantee the development of other technologies in need of regulation. Some would argue that somatic cell cloning deserves to be singled out as the most threatening possibility. But that assumes a conclusion we have not had time to reach, that Dolly-style cloning raises radically more difficult problems than, for example, cloning by embryo splitting (which can also lead to a delayed twin, with cryopreservation).[5]

NBAC admits that protecting human subjects in private research offers advantages over a ban on cloning (pp. 99–100). Yet the commission balks. It first complains that extending human subjects protections requires legislation and thus delay. But Senator John Glenn (Democrat–Ohio) has already proposed legislation,[6] and enacting a congressional ban involves delay as well. The commission further complains that human subjects legislation would rely on decentralized institutional review boards (IRBs). But others have suggested creating a national IRB for novel questions,[7] and NBAC ought to be considering this among other improvements in human subjects protection anyway. Moreover, IRBs are actually part of a larger mechanism providing centralized federal agency review when needed. The commission's final objection is that human subjects legislation would not reach beyond research activity to clinical use, as in infertility clinics. But this merely counsels supplemental regulation of those clinics.

NBAC's report, in fact, suffers from minimal consideration of infertility programs and reproductive technologies.[8] The commission acknowledges that the federal statute requiring fertility clinic reporting would seem to require reporting of cloning (p. 88).[9] But it ignores the broader issues plaguing reproductive technologies: the inadequacy of federal and state regulation, state-to-state inconsistencies, and conflicts of interest inherent in industry self-regulation. The report overlooks the burgeoning literature on those problems and, indeed, reflects little input from infertility programs.[10]

Instead of developing a legal response to cloning that addresses the core problems of private research and underregulated reproductive technologies, NBAC simply called for a ban of cloning itself. That skirts the central problems, while adding new ones.

Research Would Be Chilled by a Ban

No other bioethics controversy has been addressed by a ban as broad as the one NBAC advocates and the president now proposes. Its prohibition reaches all public and private institutions, whether or not federal money is involved or FDA approval is required. Limits on the use of federal money are common, but federal prohibitions on medical and scientific work in the private sector are not.

Moreover, the ban threatens substantial damage. The president's bill prohibits "somatic cell nuclear transfer with the intent of introducing the product of that transfer into a woman's womb or in any other way creating a human being," and would impose significant fines. Though NBAC insists it does not want to tamper with research in the private sphere, merely baby-making, this ban cannot avoid the former. The policing necessary to enforce the ban will require intruding into labs and monitoring the "intent" of scientists. Research will thus be chilled. It will be chilled further by the vagueness of a prohibition that is meant to ban baby-making, but seems to reach intent to "transfer," even if a researcher knows no child will result, plus the intent to create a human being in any unspecified "other way."

Beyond the ban's breadth and potential damage, NBAC and the president have placed this weapon in the wrong hands. The ban is to be imposed by Congress itself, not a regulatory body poised to respond to developments in the technology. That turns cloning into a political football. Past congressional brawls over the related areas of embryo research and abortion predict the same for cloning. This means that although the president and NBAC would ban private-sector application not research, Congress is likely to ban research too, as one of the pending federal bills seems to propose.[11] And though the president and commission would ban only somatic cell nuclear transfer, Congress may well include other technologies such as embryo splitting (which, after all, is another form of cloning and may also produce a delayed twin). Two of the three federal bills pending appear to do exactly that.[12] But embryo splitting may allow a woman undergoing *in vitro* fertilization to avoid repeated exposure to drugs inducing superovulation, which may reduce her risk of ovarian cancer later in life. Finally, though NBAC and the president would limit the ban to five years, there is little reason to expect Congress to develop the political bravery to lift the ban at that point.

The ban proposed thus raises serious constitutional questions. The ban's prohibition of somatic cell nuclear transfer with the wrong intent and its unavoidable chilling effect on research may infringe freedom of scientific inquiry in violation of the First Amendment.[13] And the ban as proposed by the president may well be unconstitutionally vague in its statement of the prohibited intent.[14] The ban may also represent an unconstitutional infringement on the procreative liberty of infertile couples[15] In any case, it may exceed the limits of federal power, especially since the regulation of health and clinical practice has traditionally fallen to the states.[16]

Beyond the constitutional questions, a ban at this point is bad policy. NBAC's advocacy of this ban contradicts its call for careful study and debate in our pluralistic society. With only ninety days to report on cloning, NBAC admits more analysis is needed. Yet by calling now for a ban that is likely to sweep more broadly and last much longer than NBAC wants, the commission has in effect already yielded to those who claim cloning is wrong in all cases and for the indefinite future. This ends the important deliberation, embraces one absolutist moral perspective, and writes it into law.[17]

NBAC defends the ban as a safety measure preventing harm to potential children. But that reasoning does not justify this result. Indeed, the ban may well cause harm. A ban that inevitably chills research will prevent the development of a cloning technology that is physically safe for the children it produces. Some may protest that even physically safe cloning may threaten psychological harms. But that claim is purely speculative and can ground regulation and research, but not a ban; cloning may in fact save children from psychological difficulties involved in having an anonymous genetic parent through donor egg or sperm.

Moreover, a ban may cause harm to infertile couples, especially if it hardens into an indefinite prohibition. After all, cloning offers potential benefit in infertility cases. NBAC points to a couple each carrying a recessive gene for a serious disorder. Cloning would allow them to avoid conceiving an embryo with the disorder and facing selective abortion. In another case, a woman might carry a dominant gene for a disorder. Cloning would permit her to avoid genetic contribution from an egg donor and thus would keep the genetic parenting between the woman and her partner, something of value to many couples. Other cases would include a couple entirely lacking gametes.

All of these potential uses for cloning are controversial and might ultimately be rejected. But for NBAC to ban cloning because it currently is unsafe, with no agreement on the future benefits and harms if it becomes safe, is ill-advised. Stalling development of reproductive technologies may trap us in halfway measures, such as donors' genetic involvement, that may cause more harm than cloning.

A federal ban on cloning thus misses the big picture. Cloning is only one of many reproductive technologies that should be safe before application, be it intracytoplasmic sperm injection, cytoplasm transfer, or beyond. The task is to devise a regulatory approach that addresses safety while permitting research and progress in a sphere of immense importance to couples. Cloning should spur us to that delicate balancing act. Simply lowering the boom on cloning does the opposite.

A New Advisory Body Is Required

There is a better way. Certainly we need improved regulation of assisted reproduction and human subjects experimentation in the private sphere.[18] But we have to combine that regulation with an advisory body providing oversight for cloning and other novel reproductive and genetic technologies.

The commission, president, and Congress should consider a model we have used before: agency regulation guided by an advisory body able to respond to improvements in the technology over time and more removed than Congress from partisan politics. Though NBAC's report compared policy options, strangely this was not among them.[19]

The Recombinant DNA Advisory Committee (RAC) is one example of such a body. RAC was formed over twenty years ago as an NIH advisory panel. When concern later erupted over human gene therapy, RAC (with its Working Group on Human Gene Therapy) showed how an advisory committee can hold the line, by refusing to consider germ-line gene therapy protocols for approval. It used not a legislative ban, but the committee's declared moratorium, continually subject to debate and reconsideration.

RAC's very accomplishments have fed criticism. As some forms of gene therapy became better understood (in part thanks to RAC), the committee's review began to seem an obstacle to scientific progress. The director of NIH restructured RAC earlier this year.[20] Now a smaller RAC will advise on ethical issues, surrendering authority to approve protocols to the FDA. Though RAC's authority has been reduced, this is a success story. A mechanism appropriate at the introduction of a controversial technology may require revamping later. What we use now to govern cloning must have the flexibility to evolve.

RAC is merely one example. And it is narrower than what we need for cloning: RAC's jurisdiction has been confined to protocols requiring NIH approval. On cloning, as I have argued, we need to extend human subjects protections to private research and regulate reproductive technologies, with an advisory body for novel issues such as cloning.[21]

Certainly the details of the model can be debated. Indeed, rather than create a new advisory body, using a reinvigorated RAC, another preexisting entity, or NBAC itself (if its mission were restructured) might be considered. And some may argue we need two bodies, one for human subjects and the other for reproductive technologies. But surrendering cloning to a congressional ban, as NBAC suggests, attempts a delicate operation with far too blunt an instrument. It is slim consolation that under the president's proposal, NBAC will be continuing discussion on the sidelines.

NBAC might respond that it favored a limited ban to head off worse proposals in Congress. But a national bioethics commission should call for what is right, not merely what is expedient. Congressional bills in the panicked days after the announcement of Dolly should not drive the national bioethics agenda.

A congressional ban may seem simple and safe. Yet the issues posed by cloning are not simple. We have to balance the promise of research and the potential benefits against the need for regulation and caution. We have to do better than NBAC's ban.

Footnotes to Reading 52

1. National Bioethics Advisory Commission Cloning Human Beings: Report and Recommendations of the National Bioethics Advisory Commission (Rockville, Md., June 1997).

2. The White House, Office of Communications, Directive on Cloning, 4 March 1997, 1997 Westlaw 91957 (White House).

3. See "Statement by the President on NIH Recommendation Regarding Human Embryo Research," U.S. Newswire (2 December 1994); Omnibus Consolidated Appropriations Act, 1997, Pub. L. No. 104–208, (section)512, 110 Stat. 3009, 831.

4. 45 C.F.R. Part 46 (1996). These regulations cover only research that is federally funded, at institutions offering assurances that all research will be subject to the regulations, or on drugs and devices needing FDA approval.

5. NBAC's report leaves unclear the proper policy approach to embryo splitting. Chairman Shapiro's transmittal letter states, "We do not revisit … cloning … by embryo splitting." However, a report footnote ambiguously "observes that … any other technique to create a child genetically identical to an existing … individual would raise many, if not all, of the same non-safety-related ethical concerns raised by … somatic cell nuclear transfer" (p. iii, n. 1). One would think that "any other technique" could include embryo splitting with cryopreservation to produce a delayed genetic twin. However, the report claims that the capacity to produce a delayed genetic twin is a prospect "unique" to Dolly-style cloning, i.e., somatic cell nuclear transfer (pp. 3, 64). This leaves NBAC's approach to embryo splitting in confusion.

6. S. 193, 105th Cong. (1997).

7. See Carol Levine and Arthur L. Caplan, "Beyond

Localism: A Proposal for a National Research Review Board," IRB 8, no. 2 (1986): 7–9; Alexander Morgan Capron, "An Egg Takes Flight: The Once and Future Life of the National Bioethics Advisory Commission," *Kennedy Institute of Ethics Journal* 7 (1997): 63–80, at 69.

8. NBAC might respond that its mandated areas of study are human subjects research and genetic information, not reproductive technologies. But one cannot do justice to cloning without considering its most likely use in treating infertility. And a national commission should bring to cloning the necessary bioethics analysis, not just bureaucratically designated topic areas.

9. See 42 U.S.C.A. (section) (section) 263a–1 et seq.

10. NBAC asserts, for example, that most reproductive technologies aside from *in vitro* techniques for fertilization involve no micromanipulation as substantial as somatic cell nuclear transfer (p. 32), without even analyzing techniques such as assisted hatching and cytoplasm transfer. The report also makes the startling suggestion that childlessness condemns one to immaturity: "Without reproduction one remains a child ... With reproduction ... one becomes a parent, taking on responsibilities for another that necessarily require abandoning some of the personal freedoms enjoyed before" (p. 77). An infertile adult does not automatically remain a "child" and may take on numerous responsibilities requiring self-sacrifice. Surely, these remarks would not have survived serious engagement with clinicians in infertility programs. NBAC's witness list includes none. Cf. Gina Kolata, "Ethics Panel Recommends a Ban on Human Cloning," *New York Times*, 8 June 1997, at 22 (quoting an NBAC member remarking that no IVF physicians addressed the commission).

11. See H.B. 923: "It shall be unlawful ... to use a human somatic cell for the process of producing a human clone." This seems to prohibit making even an embryo clone for research.

12. See H.B. 922, S. 368.

13. See generally Ira H. Carmen, *Cloning and the Constitution: An Inquiry into Governmental Policymaking and Genetic Experimentation* (Madison: University of Wisconsin Press, 1985); Richard Delgado and David R. Millen, "God, Galileo, and Government: Toward Constitutional Protection for Scientific Inquiry," *Washington Law Review* 53 (1978): 349–404.

14. Cf. *Lifchez v. Hartigan*, 735 F. Supp. 1381 (N.D. Ill.), aff'd without opinion, 914 F.2d 260 (7th Cir. 1990), cert. denied

sub nom. *Scholberg v. Lifchez*, 498 U.S. 1069 (1991) (striking down a statute on fetal experimentation as unconstitutionally vague).

15. The shape of this argument is suggested by John A. Robertson in *Children of Choice: Freedom and the New Reproductive Technologies* (Princeton: Princeton University Press, 1994), though he questions whether cloning is so different from other forms of reproduction as to fall outside of constitutional protection for procreative liberty (pp. 169–70). On constitutional protection for reproductive technologies, see also Lifchez, above.

16. For the limits of federal power based on the Constitution's commerce clause, see, for example, *U.S. v. Lopez*, 514 U.S. 549 (1995).

17. See also Alexander Morgan Capron, "Inside the Beltway Again: A Sheep of a Different Feather," *Kennedy Institute of Ethics Journal* 7 (1997): 171–79, at 176 ("(I)t would be a mistake to say everything we believe would be wrong to do should be a wrong to do. This is particularly true of cloning.").

18. See also George J. Annas, "Regulatory Models for Human Embryo Cloning: The Free Market, Professional Guidelines, and Government Restrictions," *Kennedy Institute of Ethics Journal* 4 (1994): 235–49, 245–46.

19. NBAC did mention RAC (p. 97), but in its discussion of voluntary moratoria (and even though RAC's moratorium on germ-line gene therapy proposals has been binding on researchers seeking federal funds, not voluntary).

20. National Institutes of Health, "Notice of Action under the NIH Guidelines for Research Involving DNA Molecules," 62 Fed. Reg. 4782 (31 January 1997).

21. Unlike a ban on cloning, my suggested approach is likely to survive constitutional scrutiny. Research is routinely disseminated interstate with substantial commercial effects. And the terrible history of research scandals would seem to justify extending protection to subjects in private research as a matter of civil and human rights. Moreover, there is little reason to suspect infringement on researchers' freedom of inquiry from application of our current protective framework. Augmenting regulation of reproductive technologies, if carefully done to respect the constitutional need for a compelling justification to restrict access to procreative technologies, would seem defensible given extensive interstate commerce in reproductive services.

54

A False Sense of Urgency Has Driven the Cloning Debate

Andrea L. Bonnicksen

Mary had a little lamb
She cloned it from a ewe
The lamb, confused (did ask) the sheep
Am I me or you?

—P. Seares and L. Seares (1997)

This apocryphal lamb's confusion is but one perplexity in the aftermath of Dolly's birth. For the first time in the era of new reproductive technologies, a research development has directed attention straight to the legislatures. The impulse to enact prohibitive laws at the front end of medicine contrasts to the usual process in which ethical and legal deliberation come first and lawmaking later, if at all. With Dolly's birth, however, a U.S. presidential decree, mobilization of a national advisory board, a dozen state bills, and three congressional bills followed with head-turning efficiency.

Intense governmental interest in a technique used successfully only once in a mammal is unprecedented. Alexander Capron's observation that "many of the issues in cloning are uncharted" holds true for policy as well as ethics (1997: 173). To be sure, there would be much to condemn in cloning an adult's genome through somatic cell nuclear transfer and very little, if anything, to commend. However, to mandate a prior restraint on a speculative technique under a false sense of urgency is to invite an unfortunate precedent that elevates politics above science and emotion above insight. A preferable alternative is to construct a cloning policy that combines private and public oversight and that incorporates two other potential methods of replicating genomes:

This appeared as "Creating A Clone in Ninety Days: In Search of a Cloning Policy" *Politics and the Life Sciences*. v16 (Sept. 1997), 1997. pp. 304–8.

twinning and embryo cell nuclear transfer.

There Was an Air of Unreality in the Cloning Debate

Twenty-four hours after Dolly's birth, President Clinton asked the newly formed National Bioethics Advisory Commission (NBAC) to conduct a "thorough review of the legal and ethical issues" associated with cloning by somatic cell nuclear transfer, to make "recommendations on possible Federal actions to prevent its abuse" (Purdum, 1997), and to report to him in ninety days. One week after Dolly's birth, the President extended the funding ban on embryo research to include research aimed at cloning human beings, and he urged a voluntary moratorium by all privately and publicly funded researchers "on the cloning of human beings until our Bioethics Advisory Commission and our entire nation have had a real chance to understand and debate the profound ethical implications of the latest advances" ("Remarks by the President on Cloning," 1997).

By all accounts, commission members worked at breakneck speed to meet the deadline, which they did after securing a three-week postponement. In the space of those three months plus three weeks, commissioners listened to or read the written testimony of at least nineteen experts, collected other material, and wrote a 107-page report that reflected consensus among the seventeen diverse members (*Cloning Human Beings*, 1997). President Clinton then announced, following the NBAC recommendations, that he would introduce legislation to the U.S. Congress forbidding the cloning of human beings and extending the moratorium until such legislation was enacted ("Remarks by the President at Announcement," 1997).

The process was impeccable, the commissioners reputable, and the intention laudatory, as at last a national-level commission existed to review issues in biomedicine. A fortunate outcome of the experience was the way it solidified the NBAC as a forum for a public, cross-disciplinary review of a troublesome research development (Capron, 1997: 179). Yet in reviewing the chronology of this executive-level response to Dolly's birth, one is tempted to travel back to the president's supplication to researchers not to clone a human being in the ninety days it would take the NBAC to sort out the ethical and legal issues of cloning.

It is this request, together with the short deadline, that injects an air of unreality in the cloning discussion and reveals the force of politics. It took Ian Wilmut and his research team 277 attempts to generate Dolly and many other attempts to generate seven other lambs through simpler kinds of nuclear transfer (1997). The science of nuclear transfer is rudimentary, albeit growing (Kolata, 1997), and that of somatic cell nuclear transfer is even more so. It was unlikely another Dolly would be generated within ninety days, let alone a cloned human being.

Moreover, in this area, a gap separates animal research and human clinical application. Whereas researchers have pursued their goal of cloning animal embryos for commercial reasons without oversight by ethicists and lawmakers, two researchers at George Washington University were severely chastized for attempting experimentally to divide (or twin) human embryos in 1993 (Hall et al., 1993). That attempt to duplicate a genome by an ethically simpler method than somatic cell cloning, and done with abnormal embryos that would have been discarded in any event, unleashed what one observer called an "ethical hullabaloo," as well as a sanctioning of the researchers by the university for not having followed required Institutional Review Board protocol (Cohen, 1994). The point is clear: even if the wherewithal to clone a human in ninety days could suddenly materialize, it would take a naive researcher to generate a conceptus and an even more naive physician to give it a try in the clinic.

If cloning a human being in ninety days was scientifically and politically untenable, why did the president ask researchers not to do it? For that matter, why did he announce that the federal government would not fund research on human cloning in light of the fact that (a) a moratorium on funding human embryo research was already underway; and (b) no proposals for human cloning were in line because somatic cell cloning in mammals was dismissed as unlikely before Dolly's birth? The president justified the moratorium by noting that a loophole possibly exists in regulations that would allow cloning. This may indeed be the case, yet one still ponders a central question: Why the rush?

One explanation is that the president was trying to allay public fears. The prospect of asexual reproduction is indeed unsettling, and public opinion polls show widespread disapproval of cloning. Nevertheless, the public fear is not so pervasive that it justifies urgent action. Citizens are fearful of war, radiation, serial killers, and airplane bombings, but "fear" may not be the best word to describe the public response to Dolly's birth. The voluminous cartoons, puns, and jokes, paired with pictures of the taciturn Dolly, suggest that in addition to anxiety—which is certainly expressed through humor—a perverse fascination percolates as well. For every Leon Kass who thoughtfully explored what he called cloning's "repugnance" (1997), there is a person who, like Steven Dawkins, opines, "I think it would be mind-bogglingly fascinating to watch a younger edition of myself growing up in the twenty-first century instead of the 1940s" (quoted in Butler and Wadman, 1997: 9).

The answer to the question, "Why the rush?" is that there was no rush, at least not of the sort that would justify feeding the mindset that quick legislation was vital. The early stages of politics were at work and they extended to the state legislatures, where a dozen anti-cloning bills materialized in the weeks following Dolly's birth. Although these bills may never actually become law, they reveal some of the pitfalls of science politics. Varied and vague, they would produce, if enacted, a patchwork quilt of hard to interpret and possibly unconstitutional laws.

An early version of a bill introduced in Florida would "(prohibit) the cloning of human DNA in the state" (Angerame, 1997); a later version modified the bill to forbid "cloning a human being," where that is defined as "creating a new individual by using the complete nuclear genetic material of an existing human being to create a second genetic duplicate of that human being" (Florida House Bill hb 1237c1, 1997). A New Jersey bill made cloning a crime, defining the cloning of a human being as "the replication of a human individual by cultivating a cell with genetic material through the egg, embryo, fetal and newborn stages into a new human individual" ("An Act concerning genetic material," 1997).

Several other states replaced earlier language with the more precise definition recommended by the American Society for Reproductive Medicine: human cloning is "the practice of cloning an existing or previously existing human being by transferring the nucleus of an adult, differentiated cell into an oocyte in which the nucleus has been removed and implanting the resulting product for gestation and subsequent birth" (Younger, 1997). A bill introduced to the U.S. Congress in early March would make it "unlawful for any person to use a human somatic cell for the process of producing a human clone." The bill did not define a "human clone," although an accompanying bill did (H.R. 923,

1997; S. 368, 1997; H.R. 922, 1997). A revised bill, reported to the House from the Committee on Science on August 1, 1997, defined "human somatic cell nuclear transfer" more precisely as "a transfer of the nucleus of a human somatic cell into an oocyte from which the nucleus has been removed or rendered inert" (U.S. Congress, 1997).

Varied penalties accompanied definitions of forbidden research in the states. California's bill would attach administrative fines for violating the proposed ban on human cloning (SB 1344, California, 1997). New Jersey's bill would criminalize cloning, and it attached a fine of U.S.$100,000 to $200,000 and a prison term of 10–20 years for violating the law ("An Act Concerning Genetic Material," 1997). North Carolina would make cloning and conspiracy to clone a felony ("A Bill to Be Entitled," 1997). Florida, Alabama, and New York would make cloning a felony of the first degree (Florida House Bill hb1237c1, 1997; Alabama S511, 1997; New York 2877–B, 1997).

The point is that while lawmakers share a sentiment to do something about cloning, the issue is too new to bring insight about what should be forbidden, why, how, or by whom. The NBAC members noted two persistent misunderstandings about cloning. The first was that through cloning, person X would beget an identical person X' without having to go through gestation, infancy, and adolescence (Bill Clinton would beget an instant second Bill Clinton). The second was that person X would beget person X' but it would take a while because person X' would need to be gestated, born, and raised (Bill Clinton would beget a baby Bill who would need to grow up first) (*Cloning Human Beings*, 1997: 2).

These ideas about identical clones untempered by the gestational and rearing environment and the mitochondrial DNA from donor eggs are not surprising given the carbon copy and xeroxing imagery that has filtered through the literature for years. And, while lawmakers may indeed be more knowledgeable about cloning, both the misperceptions among the broader public and the assorted definitions of cloning in state legislatures advise caution before lawmaking. One inconsistent and ineffective mix of state laws governing embryo research already exists in the United States; there is no need to add a second. If legislation is the answer, it should follow a model law crafted to make scientific and clinical sense, not laws quickly enacted with an undercurrent of political posturing.

Constructing A Cloning Policy

To caution against assorted anti-cloning laws in the state and federal legislatures is not to excuse somatic cell cloning. On the contrary, a powerful case can be made to resist it, perhaps eventually by legislation. Although commentators have argued that cloning may be the only way some infertile couples or those at genetic risk can conceive a child genetically related to them, it is highly questionable whether the genetic neediness of a small number of couples justifies the extraordinary societal changes that asexual reproduction would bring about. With donated eggs, donated sperm, donated embryos, surrogacy, and adoption available as alternatives for infertile people, cloning would be a luxury, not a necessity. Whatever emotional trauma might visit people who must forsake a genetic relation to their children is minor compared with the social ills that would follow were people with an aversion to uncertainty able to clone their genomes in an ultimate asexual act of self-propagation.

If efforts to thwart somatic cell cloning are thus commendable, but politically-induced anti-cloning laws are questionable, how should lawmakers proceed? Looking to the NBAC's invitation to deliberate in order "to enable society to produce appropriate long-term policies," it makes sense to look beyond short-term solutions and instead construct a policy of cloning (*Cloning Human Beings*, 1997: 110). Here "policy" is defined in its broader sense as a set of principles and norms that reflect key values and serve as a guide to action. Given the U.S. constitutional protections accorded freedom of inquiry and reproductive liberty, much of the policy might rely on existing regulations and revitalized private sector sanctions. For example, if professional associations conclude that somatic cell cloning is unethical, scientists who attempt clinical cloning might be censured or lose access to data collection procedures, such as those related to the Fertility Clinic Success Rate and Certification Act (1992).

A forward-thinking cloning policy should also recognize other methods of procreation by cloning, which raise some issues that echo those of somatic cell cloning. Procreation by cloning is defined here as generating two or more individuals who share a genome where only one would have existed before. Three methods of replicating genomes exist. The simplest is twinning (also called blastomere separation), a procedure in which technicians would remove the zona pellucida from a four- to eight-cell embryo, separate the cells (blastomeres), add an artificial zona, and allow each to cleave. Done for years with cattle, twinning by blastomere separation allows the generation of a limited number of genetically identical embryos. If done with humans in the clinical setting, it could generate genetically identical twins, triplets, or quadruplets.

Twinning replicates something that happens under natural conditions when embryos divide and identical twins, triplets, or more are born. However, clinically twinned embryos may be frozen and thawed for later transfer. This raises the prospect of spaced identical twins or triplets, whose nuclear genes would be the same but whose different gestational experience and rearing milieu would produce personality and other differences.

A second form of procreation by cloning, also untested in humans, is embryo cell nuclear transfer, in which technicians would remove the nuclei from the cells of a four- to eight-cell embryo and transfer them to enucleated donor eggs. Here the offspring would have identical nuclear DNA, but they would have different cytoplasm (and mitochondrial DNA) if different egg donors were used. Embryo cell cloning has been associated with high fetal loss and birth abnormalities in animals. Theoretically, a large number of genetically identical embryos could be generated if an expanded embryo were used, although it is difficult to visualize who would need or want large numbers of genetically identical children.

The third form of cloning is somatic cell nuclear transfer, the form that scientists dismissed as unlikely until they tried depriving differentiated somatic cells of nutrients in the process of nuclear transfer, and produced Dolly the lamb. The safety of this process is uncertain at present, and nothing about it would mimic procreation under natural conditions.

These types of genomic replication differ in several ways. First, twinning is technologically more imminent than embryo cell nuclear transfer, which in turn is more imminent than somatic cell nuclear transfer. Second, twinning and embryo cell nuclear transfer would be performed primarily to generate more embryos to transfer during infertility treatment, while somatic cell nuclear transfer might be done for reasons other than circumventing infertility (e.g., allowing a person to procreate without a partner). Third, and perhaps most important, somatic cell cloning is the only type of procreation by cloning in which a genome in replicated without the intermixing of genes between a male and female. Even though the clone would not be phenotypically identical to his or her progenitor, procreation by somatic cell cloning would produce more predictable children than procreation by the usual genetic lottery.

Despite the differences, it makes sense to consider all cloning methods when developing principles for private and public rulemaking. The three methods touch upon common issues such as the emotional well-being and individuality of the offspring, safety of the procedure, the question of need, and the emphasis each places on genetic relationships. Because of these common threads, it is illogical to look at one and not the others. The multiple commentaries published after the ill-fated twinning experiment at George Washington University in 1993 show that embryo replication touches a sensitive chord. In fact, many of the arguments against somatic cell cloning—such as threats to individuality—were first carefully articulated in response to the twinning study (see, e.g., "Ethics and the Cloning," 1994). If these arguments are to be taken seriously, it would be odd for somatic cell cloning to be treated separately from embryo cloning and twinning.

A comprehensive cloning policy would cover all three types of cloning—twinning, embryo cell nuclear transfer, and somatic cell nuclear transfer. It would disallow all forms, allow all forms, or disallow some forms and allow others. When allowed, it would also specify the conditions under which each type might be used. The implementation might be by public law, private policymaking, or a mixture of the two. The aim is not to ban methods of cloning by isolated law, but instead to identify common and divergent issues of cloning, develop a graduated policy, and integrate sanctions into the existing oversight apparatus in laboratory and clinical research.

It may indeed turn out that only somatic cell nuclear transfer should be forbidden. By reaching this result through careful policy building rather than through preemptive and draconian laws, the outcome is more likely to have legitimacy. In addition, it will not establish as stark a precedent for prior restraints on science and medicine, and it will present the opportunity to consider the merits and demerits of twinning and embryo cloning, and thus avoid the risk that both may seem innocuous when compared with somatic cell cloning.

In addition, a careful policy-building approach would help move attention away from the speculative element of cloning and toward more realistic scenarios of clinical use. In their simplest forms, twinning and embryo cell cloning would be used by couples who are employing *in vitro* fertilization (IVF) and who have only one or two embryos and want to maximize their chances of conception by dividing the embryos to produce more for transfer. In a more complex variation that raises more issues, a couple would freeze some embryos and give birth to spaced twins or triplets. In still another variation, the couple would have spaced twins and then donate the third embryo to another couple, which would lead to spaced twins who have different gestational mothers and are living with different families.

In its simplest form, somatic cell nuclear transfer would be used by a couple, both of whom are infertile, in order to have a child genetically related to one of them (Robertson, 1997). Yet the odds that cloning would be limited to needy infertile couples are quite low. More likely, cloning would appeal to hearty egos intent on perpetuating themselves without sharing their genes with a partner, aging individuals facing their own mortality who would hire a surrogate to bear their genetic offspring, or couples seeking a child with specific traits.

Differentiating the three forms of cloning helps produce a richer understanding of the motives that might prompt their use. It also suggests more refined questions. For example, what is it about each type of cloning that is disturbing? What, if anything, about each type should be limited? If limits are advisable, upon what grounds will those limits be justified? What will be the enforcement mechanisms? Will these mechanisms comport with constitutional protections?

Science fiction scenarios need to give way to examinations of the essential homeliness of cloning in the clinical setting. This will allow us to understand who would try cloning and why, to anticipate likely justifications, and to arrive at workable policy schemes. The NBAC had it right; the discussion must continue, yet this discussion need not culminate in federal anti-cloning legislation at this early date. It is hard to imagine any legitimate use of somatic cell cloning, given the myriad alternatives in assisted conception that allow couples to reproduce. Yet the distasteful quality of somatic cell cloning is the very thing that should motivate a well-crafted and multi-hued approach to cloning that will not be threatened by early obsolescence. A ban can always be lifted; a policy based on expectations and norms is not so easy to uproot.

Acknowledgments

A similar version of this commentary will appear in a symposium issue on cloning in *Jurimetrics*, November, 1997. I thank the editors of Jurimetrics for granting permission to publish this essay here as well.

References

"An Act Concerning Genetic Material and Information, Supplementing Title 2C of the New Jersey Statutes and Amending P.L. 1996, c. 126." (1997), March 24.

Alabama S511. "A Bill to Be Entitled an Act to Prohibit the Cloning of Human Beings." (1997). March 4.

Angerame, L. (1997). "Cloning Captivates American Public; Inspires Spate of Federal and State Hearings, Legislation." *ASRM News* 31 (Summer): 22.

"A Bill to be Entitled 'An Act to Ban the Cloning of a Human Being " S. 782 (1997). North Carolina, April 9.

Butler, D. and M. Wadman (1997). "Calls for Cloning Ban Sell Science Short." *Nature* 386 (March 6): 8–9.

Capron, A.M. (1997). "Inside the Beltway Again: A Sheep of a Different Feather." *Kennedy Institute of Ethics Journal* 7: 171–79.

Cloning Human Beings: Report and Recommendations of the National Bioethics Advisory Commission (1997). Rockville, MD. June.

Cohen, C.B. (1994). "Future Directions for Human Cloning by Embryo Splitting: After the Hullabaloo." *Kennedy Institute of Ethics Journal* 4: 187–92.

"Ethics and the Cloning of Human Embryos" (1994). Special Issue *Kennedy Institute of Ethics Journal* 4 (3).

Fertility Clinic Success Rate and Certification Act of 1992. 106 Stat. 3146, P.L. 102–493.

Florida House Bill hb 1237c1 (1997). March 7.

H.R. 922. "A Bill to Prohibit the Expenditure of Federal

Funds to Conduct or Support Research on the Cloning of Humans" (1997). 105th Cong., 1st Sess. March 5.

H.R. 923. "A Bill to Prohibit the Cloning of Humans" (1997). 105th Cong. 1st Sess., March 5.

Hall, J.L., D. Engel, P.R. Gindoff, G.L. Mottla, and R.J. Stillman (1993). "Experimental Cloning of Human Polyploid Embryos Using an Artificial Zona Pellucida." Paper presented at the 1993 Annual Meeting of the American Fertility Society. Program Supplement.

Kass, L.R. (1997). "The Wisdom of Repugnance." *New Republic* (June 2): 17–26.

Kolata, G. (1997). "Rush Is Underway for Cloning of Animals." *New York Times* (June 3): B9.

New York 2877–B, "An Act to Amend the Public Health Law, the Penal Law and the Agriculture and Markets Law, in Relation to Prohibiting the Cloning of a Human Being" (1997).

Purdum, T.S. (1997). "President Asks Experts for Advice on the New Reality of Cloning." *New York Times* (February 25).

"Remarks by the President at Announcement of Cloning Legislation" (1997). Office of the Press Secretary. The White House, June 9.

"Remarks by the President on Cloning" (1997). Office of the Press Secretary. The White House, March 4.

Robertson, J.A. (1997). Testimony before the Subcommittee on Public Health and Safety. Committee on Labor and Human Resources. U.S. Senate. June 17.

S. 368. "A Bill to Prohibit the Use of Federal Funds for Human Cloning Research" (1997). 105th Cong. 1st Sess. February 27.

SB 1344. California (1997). March 11.

Seares, P. and L. Seares (1997). "Metropolitan Diary." *New York Times* (March 16).

U.S. Congress (1997). "Human Cloning Research Prohibition Act." House Report 105–239, Part 1. 105 Cong., 1st Sess., August 1.

Wilmut, I., A.E. Schnieke, J. McWhir, A.J. Kind, and K.H.S. Campbell (1997). "Viable Offspring Derived from Fetal and Adult Mammalian Cells." *Nature* 385 (February 27): 810–13.

Younger, J.B. (1997). Executive Director, American Society for Reproductive Medicine, Letter to Senator John H. Carrington, April 15.

Index